Selfie

Also by Will Storr

NON-FICTION

The Unpersuadables: Adventures with the Enemies of Science

Will Storr Versus the Supernatural

FICTION

The Hunger and the Howling of Killian Lone

Selfie

How We Became So Self-Obsessed and What It's Doing to Us

Will Storr

The Overlook Press
New York, NY

This edition first published in hardcover in the United States in 2018 by
The Overlook Press, Peter Mayer Publishers, Inc.

141 Wooster Street
New York, NY 10012
www.overlookpress.com

For bulk and special sales, please contact sales@overlookny.com,
or write us at the address above.

Cataloging-in-Publication Data is available from the Library of Congress

Manufactured in the United States of America
ISBN: 978-1-4683-1589-9

1 3 5 7 9 10 8 6 4 2

For Charles Whitman,
who was right.

“Day or night, it’s always true,
The kingdom lives inside of you.
When you say these words, three times in a row:
‘I’m lovable, I’m lovable, I’m lovable!’
Your lovable self will magically grow.”

Diane Loomans,
The Lovables in the Kingdom of Self-Esteem (1991)

"I am done with the monster of 'we,' the word of serfdom, of plunder, of misery, falsehood and shame.

And now I see the face of god, and I raise this god over the earth, this god whom men have sought since men came into being, this god who will grant them joy and peace and pride.

This god, this one word:

'I.'

Ayn Rand, *Anthem* (1931)

Contents

A note on the text

A significant part of this book concerns differences between groups of people. Sometimes generations are compared, at other times cultures. It's important to stress that these are always general tendencies that academics have detected across large numbers of people. In the real world, there is a huge variation amongst individuals, and no general observation about a specific group can ever be reduced to an observation about any particular person.

BOOK ZERO

The Dying Self

At first there was nothing. She was a person, tied to a bed, and that was all. No memories, no thoughts, only strange sounds: electronic beeps, a soft mechanical drone. And then, emerging from the haze, a voice: 'Can you tell me what this is?'

Something was floating in front of her.

'A pen?' she said.

In gauzy glimpses of awareness she'd recognize shapes in the room – a bed, a chair – but somehow she was unable to turn all the individual pieces into a whole, connected scene. Human forms would stand over her and tell her they knew her, but she had no idea who she was. She didn't know it was the second week of June 2007 or that she was forty-three years old or that she was Debbie Hampton from Greensboro, North Carolina. At some point, however, she did come to understand the fundamental facts of her predicament. She was alive. And she was furious about it.

A few days previously, Debbie had taken an overdose of over ninety pills, a combination of ten different prescription drugs, some of which she'd stolen from a neighbour's bedside cabinet. Ever since she'd been the gangly girl that everyone at school called Monkey, Debbie had suffered from low self-esteem. Her childhood had not been easy. 'My parents got divorced when I was sixteen,' she told me, 'and I vowed, then and there, that I would never do that to my children.'

When she was twenty-one she married her childhood sweetheart and they quickly had children. She was determined to become the woman of his dreams. 'My husband's mom was the epitome of the perfect mother and wife. She stayed at home, raised the kids, was a wonderful cook, was crafty. She was what I wanted to be.' But as hard as she tried, Debbie couldn't be that woman. The life of the housewife bored her. 'I wasn't a pleasant person to be around. I was angry.' The marriage ended. And so there she was, the single mother she swore she'd never be. She began dating, but it didn't go well. 'I literally saw my youngest son sit in the middle of the hall crying big tears because he wanted a "real" dad.' As a child, Debbie had always tried to be the person her mother wanted her to be. As an adult, she'd struggled to be the person she imagined her husband desired. All her life she'd been chasing that dream of perfection and, all her life, her dream of perfection had dodged her grasp. What she felt like now was a failure. 'My thoughts were, "You're not a good enough mother, you'll never be able to earn money, you're getting older, you'll never get a man and please him."'

On 6 June 2007, at around eleven o'clock, Debbie sat on her bed, swallowed her pills with some cheap Shiraz and put on a Dido CD to listen to as she died. Rousing a little later, she went downstairs to her computer, switched it on, and began composing a suicide note:

'*Dear Family, I write this as tears fall from my eyes. I fear that I will never see any of you again. I have been captured by the white men, and I am on a slave ship headed to a far away land.*'

She really did feel extremely odd. She nevertheless continued writing for a while, and then signed off:

'*Goodbye forever and be careful not to get caught as I did, Kunta Kinte.*'

At around three that afternoon, one of Debbie's sons found her collapsed on the kitchen floor. She was rushed to hospital where she eventually awoke, raging at herself. 'I was mad,' she said. 'I'd messed it up. I was full of self-loathing for botching my suicide. I had so much self-loathing I felt like it would crush me.'

*

Suicide is a mystery. It seems to go against everything we know about human nature in some elemental way. We're animals of progress. We're doers, strivers, fighters. Whether our aims are good or ill, we push and we push, building great cities, burrowing great mines, forging great empires, destroying climates and habitats and the limits of yesterday's fantasies, bending the forces of the universe to turn magic into the everyday. We want things and we get them; we're greedy, ambitious, canny, relentless. Self-destruction has no place in this schema. It doesn't fit.

Except it does. It must do.

I wondered if there was a clue, in Debbie's story, about what it might be that can make the human self malfunction so utterly. Over the last few years, I've spoken with many people who've been affected by suicide and the basic narrative she describes, of high expectations leading to failure, leading to a rejection of the self and an impulse to finish it all off, has emerged again and again. It emerged when I spoke with Graeme Cowen, from New South Wales, who always felt that 'if I wasn't outstanding, I was less of a person' and, following a series of professional failures, tried to hang himself in his backyard with an electrical cord. It emerged when I spoke to Drummond Carter, an ambitious and proper headmaster from a village in Norfolk, England, whose wife's serial affairs left him humiliated, his identity shattered. It emerged when I spoke with star footballer Ben Ross, who broke his neck at the peak of his career ('You start thinking, "What if I was to disappear?"'), and his sports medic, Dr Con Mitropoulos, who told me his charges often find themselves thinking like this because they 'put themselves under pressure and have aspirations to be successful, as we all do. They feel anything's possible, as long as you work hard. And yet life's not like that.' It emerged, too, when I spoke with Meredith Simon, a student at a well-known US liberal arts school who struggled with her weight and with ADHD and felt significantly less perfect than her beautiful, slim and accomplished sisters. She began self-harming and then, when she was fourteen, went into her en-suite bathroom, took a razor blade and tried to kill herself by slashing her wrists. 'I was disappointing people,' she told me. 'And that was really hard for me, because I wanted to be the perfect child.'

Perhaps I notice this pattern in particular because I find it in myself. I seem to be caught in a lifelong rhythm of expecting more from myself than my talent and character can supply, and in periods of mounting failure, suicide tends to be my mind's reflexive solution. *Fuck it,* I find myself thinking, *I could just leave,* and then I get this sense of warm reassurance. I've always secretly admired those who have the courage to actually go through with it. To me, they're heroes. After all, there's not much cowardly about taking to your veins with razor blades or coiling an electrical cord around your throat.

One way of looking at suicide is as a catastrophic breakdown in the human self. It's the most extreme form of self-harm there is. Even if you haven't actively plotted your own death, many people have surely experienced at least a fleeting thought: *I could solve this. I could vanish.* I have reason to suspect that this kind of thinking, although taboo, might be more common than you might imagine. In the three years it's taken me to write this book, four self-inflicted deaths have taken place in my general sphere. There's the man who hanged himself from a tree on the common where I walk my dogs; the man who hanged himself in the lock-up garage that I pass when driving my wife to work; the lovely village postman, Andy, who I used to see almost every day, who also hanged himself; and then there's my cousin, who took his life this past Christmas.

You might argue that this is just bad luck, or that I notice suicides more because I have that vulnerability or because I happen to have been writing about them. Indeed, it's true that overall rates in the US and UK have seen a general decline since the 1980s. But it's also true that the 1980s saw the introduction of 'blockbuster' antidepressants that are known to be especially effective for severe depression, the type most likely to lead to suicide. Prescription numbers have rocketed since the early 2000s. Today, over a twelve-month period, between 8 and 10 per cent of the entire adult population of the US and UK uses antidepressants. There exists the strong possibility that our suicide statistics might look considerably worse if the millions suffering from serious psychological maladies hadn't been offered this help.

And those statistics are pretty dire already. Today more people die by suicide than in all the wars, terrorist attacks, murders and government executions combined. According to the World Health Organisation, in 2012, 11.4 people out of every 100,000 died by self-harm versus 8.8 people as a result of interpersonal violence, collective violence and legal intervention. Its projections indicate things are going to get worse. By 2030, it estimates that that difference will have increased to 12 versus 7. In the UK, in 2000, 3.8 per cent of adults reported suicidal thoughts – a figure that had jumped to 5.4 per cent by 2014. In the US, suicide rates recently hit a thirty-year high. Between 2008 and 2015, the number of American adolescents and children receiving hospital treatment after considering suicide or self harm doubled.

And as alarming as these figures are they do, in fact, heavily disguise the problem's true weight. The data varies, but one respectable set has it there are *twenty times* more attempted than completed suicides every year. That's a massive amount of people whose supposedly self-interested selves are, for some reason, turning against them. This is deeply strange. What is it that holds such incredible power it can make the human psychological mechanism go dark? That's so energetically harmful it can cause it to want to destroy itself? Could it be, I began to wonder, something to do with that pattern I'd detected, in myself and others? Something about high expectations and then disappointment and then a terrible, gathering loathing of the self?

*

'Did you see the news?' said Professor Rory O'Connor, President of the International Academy of Suicide Research, when I met him at his office. Rory heads up the University of Glasgow's Suicide Behaviour Research Laboratory. The day we met, the British papers had been carrying the latest statistics. Whilst the level for women had remained pretty constant for several years, the numbers for men were at their highest in well over a decade. As it is, men make up around 80 per cent of all suicides in the English-speaking nations, but this new surge was worrying. The papers had been asking after its causes.

We began by discussing the more general facts. Those who study suicide are usually keen to press upon the curious that there's rarely a single factor that leads to any self-inflicted death. There are many vulnerabilities that can heighten risk, including impulsivity, brooding rumination, low serotonin and poor social-problem-solving abilities. Mental illness, most commonly depression, usually precedes such an event. 'But the really important point is, most people with depression don't kill themselves,' Rory told me. 'Less than 5 per cent do. So mental illness is not an explanation. For me, the decision to kill yourself is a psychological phenomenon. What we're trying to do in the lab here is understand the psychology of the suicidal mind.'

At forty-three, Rory was boyish in frame and youthful in spirit. Hyperactive and assertive, the cuffs of his giraffe-decorated shirt flapped open and his grey hair was fashionably cut, parted long across the brow. Paintings by his children were stuck to a cork board – an orange crab, a red telephone – whilst, all the while, a grim book collection lurked in his cupboard: *Comprehending Suicide*; *An Unquiet Mind*; *By Their Own Young Hands*.

After two decades of studying them, you'd imagine there wasn't much Rory didn't know about the minds of the suicidal. And yet, every now and then, he came across a finding that surprised him. This is just what happened when he began looking at a style of thinking called social perfectionism. If you're prone to social perfectionism, your self-esteem will be dangerously dependent on keeping the roles and responsibilities you believe you have. You'll tend to agree with statements such as, 'People expect nothing less than perfection from me' and 'Success means that I must work harder to please others.' It's not about what you expect of yourself. 'It's what you think other people expect,' Rory explained. 'You've let others down because you've failed to be a good father or a good brother – whatever it is.'

He first came across this type of perfectionism in studies of American university students. 'I thought it wouldn't be applicable in a UK context and it certainly wouldn't be applicable to people from really difficult backgrounds,' he said. 'Well, it is. It's a remarkably robust effect. We've found this relationship between social perfectionism and suicidality in all populations where we've done the work, including among the disadvantaged and the affluent.' What's not yet known is why. 'Our hypothesis is that social perfectionists are much more sensitive to signals of failure in the environment.'

I wondered if this might be relevant to the problem of male suicides. 'If this is about perceived failure to fulfil roles, should we be asking what roles men feel they should fill? Father? Bread-winner?'

'And now there's this change in society,' he said. 'You have to be Mr Metrosexual too. There are all these greater expectations. More opportunities for men to feel like failures.'

Studies suggest it's fairly easy to make a man feel this way. One examination of what both women and men believe it takes to 'be a man' these days found they have to be a 'fighter', a 'winner', a 'provider', a 'protector' and 'maintain mastery and control' at all times. 'If you break any of those rules you're not a man,' the paper's author, clinical psychologist Martin Seager, told me. As well as all this, 'real men' aren't supposed to show vulnerability. 'A man who's needing help is seen as a figure of fun.' His was a relatively small study, but it echoed, to a remarkable degree, a report on male suicide that Rory co-authored for the Samaritans: 'Men compare themselves against a masculine "gold standard" which prizes power, control and invincibility. When men believe they are not meeting this standard, they feel a sense of shame and defeat.'

As we chatted, Rory told me about a close female friend who killed herself in 2008. 'That really had a huge impact on me,' he said. 'I kept thinking, "Why didn't I spot it? God, I've been doing this for years." I felt like a failure, that I'd failed her and people around her.'

'Sounds like social perfectionism,' I said.

'Oh, I'm definitely social perfectionistic,' he said. 'I'm hyper-sensitive to social criticism, even though I hide it well. I'm really sensitive to the idea I've let other people down.' Another risky trait he admits to suffering is brooding rumination – continual thoughts about thoughts. 'I'm a brooding ruminator and social perfectionist, aye, without a doubt. When you leave I'll spend the rest of tonight, and when I'm going to sleep, thinking, "Oh, Jeez, I don't believe I said that. I'll kill – " he stopped himself – "I'll beat myself up."'

'Rory,' I said. 'Are you at risk of suicide?'

He paused, apparently feeling out what he was about to say. 'I would never say never,' he said. 'I think everybody has fleeting thoughts at some stage. Well, not everybody. There's evidence that lots of people do. But I've never been depressed or actively suicidal, thank God.'

Rory and his team have developed a model for suicidal thinking that is based, in part, on an influential paper by the eminent psychologist Professor Roy Baumeister, in which it's described as an 'escape from the self'. Baumeister theorized that the process starts when events in a person's life 'fall severely short of standards and expectations'. The self then blames itself for these failures, and loses faith in its ability to repair what's gone wrong. 'We believe it's a feeling of being defeated and humiliated from which you cannot escape,' said

Rory. It's not enough just to feel like a failure, a self must also lose faith in its capacity to change. 'It can be internally and also externally, so you're trapped by life circumstances, you can't see a way out, or your job prospects aren't going to change and so on.'

'It's a feeling of being stuck,' I said.

'Absolutely. This sense of entrapment, which all comes back to control.'

'Control?'

'If you look at psychological theories across both physical health and mental health, one of the threads that runs through that is a sense of control. Whenever that control is breached it's really problematic. When we're distressed we're always trying to get control back.'

One of the most critical functions of the human self is to make us feel in control of our lives. When people are having perfectionistic thoughts, they're wanting to feel that they're in control of their mission of being the great person they imagine they ought to be. The problem comes when the mission's progress stops or, worse, goes into reverse. When their plans go badly awry, they'll strive to get that control back. If they fail and keep on failing, they'll enter despair. The self will begin to founder.

And this is true for all of us. If a great deal of men are suffering under the powerful cultural expectation that they must be invulnerable fighters, protectors and winners, then women have their own universe of pressures to cope with. Although the worldwide data is relatively scant, it seems that in many countries women actually attempt suicide in greater numbers. This indicates the massive pressures many feel to measure up to their own frequently impossible standards of perfection. Indeed, the assault on the female self can seem nearly constant, from the expectation that one should 'have it all', the perfect career and family (both men and women, it seems, are expected to embody the finest features of both traditional gender roles;

strong yet caring, ambitious yet family-minded – all must be globally perfect), to the sick images of 'ideal' body-shape they're presented with in fashion magazines and popular clothing stores, some of which display 'triple zero' mannequins with hazardous waist-to-height ratios of 0.32. One recent survey found that only 61 per cent of young women and girls in the UK felt happy with their bodies, a significant decline from the 73 per cent that had been found just five years before. Meanwhile, nearly a quarter of seven- to ten-year-olds felt they 'needed to be perfect', an already troubling figure that grew much worse with adolescence: in eleven- to twenty-one-year-olds, the proportion soared to 61 per cent.

A truer picture of today's burdens on the female self can only be glimpsed by widening our scope from suicide to include self-harm and eating disorders, conditions which disproportionately affect women and which also, significantly, have perfectionism as a predictor. And, when we do, it begins to look even more as if something bad is happening. Since the emergence of social media, the incidence of eating disorders and body dysmorphia in the US and the UK has risen by around 30 per cent. In Britain, the number of adults reporting self-harm between 2000 and 2014 has more than doubled. One senior psychiatrist told reporters that rises in youth self-cutting seemed to confirm the general experience of clinicians that 'levels of distress are rising . . . and mental health disorders are rising, for both boys and girls'. In the US anxiety and depression have been rising in adolescents since 2012. The authoritative American Freshman Survey found 51 per cent more young students felt 'overwhelmed' in 2016 than did in 2009. Even more worrying, the numbers for feeling 'depressed' had risen by 95 per cent.

There's likely to be a variety of causes for these jumps, possibly including improved detection or more patients feeling able to disclose their behaviours. But specialists close to the problems also point to the 'unprecedented social pressures' that young people are currently under. Dr Jackie Cornish, of

NHS England, said, 'In common with most experts, we believe this is due to increasing stress and social pressure on young people, including to succeed in school, and emerging problems with body image.' Paediatrician Dr Colin Michie placed much of the blame on the modern ubiquity of smartphones, which expose the young to constant streams of advertising and celebrity, telling reporters, 'I think we have released a behemoth we cannot control.'

Traditionally, body image troubles have been thought of as affecting mostly women. This, too, is changing. One US study found body dysmorphic disorder to be nearly as prevalent amongst men as it is amongst women. Many of these men have 'muscle dysmorphia', a condition that was essentially unheard of twenty-five years ago. During this period, steroid use has soared. Before the 1980s, it was mostly associated with a tiny number of elite athletes, whilst today it's thought that up to four million Americans, the vast majority of them men, have used the muscle-enhancing drugs. Britain has experienced an incredible 43 per cent jump in male referrals for eating disorders over the course of just two years. Meanwhile, needle exchanges in some cities have reported a 600 per cent increase in their use in the decade to 2015. Spending on gym membership in 2015 alone rose by 44 per cent. A government enquiry into the problem concluded, 'body image dissatisfaction is high and on the increase'. All of us, male and female, are apparently feeling increasing pressure to be perfect.

These dangers have also been becoming apparent in the university system. When a University of Pennsylvania task force published their report into the problem of student suicide, they noted a dangerous 'perception that one has to be perfect in every academic, cocurricular and social endeavor.' Over at York University in Toronto, meanwhile, the social scientist Professor Gordon Flett has heard the dangerous sirens of perfectionism in many of his students. 'During advising appointments, every second student seemed to be coming in with what, at first, I thought was an academic problem but was actually an emotional or coping problem related to not meeting expectations,' he told me. 'What I was seeing was exceptionally capable people swamped with anxiety and stress due to the incredible pressure of trying to be perfect, either because they'd embraced these standards as their own or, in the case of social perfectionism, because they felt that others were demanding it of them.'

Gordon co-authored a paper that argued perfectionism has been significantly underestimated as an 'amplifier' for suicidal ideation. Amongst other studies, he referenced a survey in which 56 per cent of friends and family members of someone who'd killed themselves referred to the deceased, unprompted, as a 'perfectionist'. In another, interviews with mothers of male suicide victims found 71 per cent of them saying their sons had placed 'exceedingly high' demands on themselves. The most comprehensive study on the issue to date found a 'strong link' between perfectionism and suicide. Authors of the 2017 meta-analysis described perfectionists as being 'locked in an endless loop of self-defeating over-striving in which each new task is another opportunity for hard self-rebuke, disappointment and failure.'

Of course, perfectionism isn't something we either have or don't have: it's not a virus or a broken bone. It's a pattern of thinking. Everyone sits somewhere on the perfectionist scale. We're all more or less perfectionist, with those in the upper levels being more sensitive to signals of failure in the environment. Even if you don't consider yourself a perfectionist, it's likely you have an idea of the person you feel you ought to be and experience at least a pang when you don't measure up to it. That resonant moment of longing sorrow when you realize you've failed – that's what we're talking about.

Gordon observes that it's common, these days, for people to consider perfectionism 'as an ideal', but these darker effects are lesser known, as are its shifting guises. 'Self-oriented perfectionism' is the one that isn't social – the demands for perfection come from within the self. There's 'narcissistic perfectionism', in which people believe they're absolutely capable of reaching the highest heights but become vulnerable when they finally realize that, actually, they're not. And then there's 'neurotic perfectionism', the category into which both Debbie and I probably fit. These people suffer from low self-esteem and 'just feel like they never measure up'. They're worried and anxious people who have a 'massive discrepancy' between who they are and who they want to be. They make sweeping generalizations about themselves, so if they're 'not efficient' at a particular thing, they experience it as a failure of the entire self. 'It's this all or nothing thinking,' he says. 'With that comes a lot of self-loathing.' Often, it begins with a simple belief that they don't matter, 'but if they just achieve at a certain level, they will matter. Being perfect will either compensate for these defects or fool others into thinking they don't have them.'

For Gordon, one central cause is that our environment is changing. Closely echoing Rory's observation about modern masculinity, he says the modern world is giving us a greater number of opportunities to feel like failures. 'It's something that's becoming more salient,' he said. 'In part, that's because of the internet and social media. First, when a public figure makes a mistake there seems to be a much stronger, more intense and quicker backlash. So kids growing up now see what happens to people who make a mistake and they're very fearful of it.' This seems to be what happened, for instance, in July 2016, when sixteen-year-old Phoebe Connop took her own life after becoming worried a joke photograph she'd taken of herself would lead to her being denounced as a racist. 'The image had circulated further than she wanted it to,' Detective Sergeant Katherine Tomkins told her inquest. 'There had been some negative reaction.'

Gordon and his colleagues have also recently begun studying a phenomenon called 'perfectionist presentation'. 'That's the tendency to put on a false front of seeming perfect, where you cover up mistakes and shortcomings,' he said. 'You'll see this especially among younger people, who portray their lives on social media. For the person who feels they need to keep up with others, that seems to be an added pressure. It's like, "Here's my perfect life, take a look at it".' Everybody judges their own worth by comparing themselves to others. That's simply how minds work. Because of this, Gordon believes, social media is having 'a huge effect' on people's self-image.

It's not just Gordon who thinks so. A *New York Times* report into rising suicide rates among fifteen- to twenty-four-year-olds in the US featured the Director of Counselling and Psychological Services at Cornell University, Gregory Eells, and described his belief that social media is a 'huge contributor to the misperception among students that peers aren't also struggling.' When students in counselling remark that everyone else on campus looks happy, he tells them, 'I walk around and think, "That one's gone to the hospital. That person has an eating disorder. That student just went on antidepressants."'

I've heard a similar sentiment from a friend who's a mental health nurse and talks of a startling rise in cases of what she and her colleagues now describe as 'chronic dissatisfaction'. I've also heard it from my own interviewees. Take Meredith, the liberal arts student we met earlier, who slashed her wrists in her bathroom. She believes that social media was a significant factor in her attempted suicide. 'I've grown up in an age where social media has gotten really popular,' she told me. 'When I was younger we had AOL Instant Messenger and it was how many friends you had that classified you as cool. Then Facebook became more popular. That became a challenge for me because seeing everything that was posted on there was difficult.'

'What was?' I asked.

'Just seeing people really happy. It was kind of in your face. When people were like, "Oh, I love my life," it was hard because I wanted to feel that way and I didn't. But I also posted things like that so I could seem that way.'

And then, in 2018, a major new study added significant empirical weight to the hunch I'd been pursuing for years. Psychologists analysed data from over 40,000 university students across the US, UK and Canada and found levels of perfectionism, between 1989 and 2016, had risen substantially. Over that period, the extent to which people attached 'an irrational importance to being perfect' had gone up by 10 per cent. Meanwhile, the extent to which they felt they had to 'display perfection to secure approval' had grown by a startling 33 per cent. These findings, the researchers concluded, indicated that 'recent generations of young people perceive that others are more demanding of them, are more demanding of others, and are more demanding of themselves.' It wasn't only our environment that was changing. We were, too.

*

We're living in an age of perfectionism, and perfection is the idea that kills. Whether it's social media or pressure to be the impossibly 'perfect' twenty-first-century iterations of ourselves, or pressure to have the perfect body, or pressure to be successful in our careers, or any of the other myriad ways in which we place overly high expectations on ourselves and other people, we're creating a psychological environment that's toxic. Which is not to say that perfectionism is the *only* problem or that it's unique to our era. There are many routes to suicide and self-harm just as every generation of people have surely had unfair expectations placed upon them. But if Rory and Gordon and the others are right, there are aspects of today's cultural environment that can be especially hazardous and are transforming who we are.

I want to find out how this has happened. To do so, I'm going to have to embark upon two separate investigations. First, I'll need to examine the self – the mechanism of wills, beliefs and personal qualities that combine to make us who we are, because the self is the thing that's somehow becoming changed and damaged. Every self is different, of course, but I'm going to burrow beneath this crust and discover the fundamentals of how the human self operates. We feel its power: it's the self that presses us into preoccupation with status, attraction, achievement, morality, punishment and perfection. We feel that it's uniquely 'us' that gets involved in conflicts and loves and dreams, but that all humans repeat the same patterns of behaviour betrays the fact there are actually laws and functions in operation, an apparatus of being. This apparatus began its formation millions of years ago. In tracing its journey, and mapping its design, I want to find out why perfectionistic thoughts can cause it to malfunction so badly that it would rather eat itself than survive.

My second investigation will be into the environment that's making us more perfectionistic. This means examining culture. When people feel like failures, they're comparing their own self to an ideal of what a self *should* be like and then concluding they somehow come up short. It's our culture that largely (although not entirely) defines who this ideal self is and what it looks like. It assails us with this perfect self that we're all supposed to want to be like, in films, books, shop windows, newspapers, advertising, on the television and the internet – everywhere it can. Most of us feel pressured, in some way, into living up to this cultural model of perfection.

Of course, everyone has a slightly different version of the ideal self that they're aiming to be, depending on their gender, their spiritual beliefs, their age, their family and peer background, their job and so on. Debbie's model of 'the perfect wife and mother', for example, seems to have its roots in a cultural era that will, to many, seem antiquated. It's not difficult, though, to detect the general model of ideal selfhood that the culture of today has come up with. It's usually depicted as an extroverted, slim, beautiful, individualistic, optimistic, hard-working, socially aware yet high-self-esteeming global citizen with entrepreneurial guile and a selfie camera. It enjoys thinking it's in some way unique, that it's trying to 'make the world a better place', and one of the traits it'll value highly is that of personal authenticity, or 'being real'. It'll preach that in order to find happiness and success, you must be 'true to yourself' and 'follow your dreams'. And if you dream big enough, in the words of the sports medic Dr Con Mitropoulos – who has witnessed for himself the brutal underside of some of these ideas – you'll discover that 'anything is possible'. Oh, and it's usually younger than thirty.

Because of its ubiquity in our culture, we can lose sight of the fact that this model of the ideal self is extremely odd. Who *is* this person? I'm going to try to find out, by uncovering the remarkable tale of how it came to be. Although its basic outlines predate human civilization, the journey of our current Western model of self really began in Ancient Greece. It was there that our idea of ourselves as potentially perfectible individuals who are responsible for our own fates came into being. I want to track this idea of 'Individualism' as it evolved through the ages of Christianity, industry, science and psychology, right through to Silicon Valley and the era of hyper-individualistic and competitive neoliberalism most of us have grown up in, with all the new opportunities it's brought to make us feel like failures.

Each one of the eras I've chosen to touch down in has changed us in a different way, to a degree that I find extraordinary. We often fail to realize that the things we believe are, to a significant extent, a combination of beliefs, stories, philosophies, superstitions, lies, mistakes, and struggles of flawed men and women – our culture, in a word. Voices from long-dead minds haunt us in the present, often without our conscious awareness. Arguments they've made, feuds they've waged, battles they've fought, bestsellers they've written, revolutions they've triggered, industries and movements they've raised and destroyed, all live on within us. If I'm going to understand the age of perfectionism, and this peculiar model of self that ours has somehow arrived at, I'm going to have to tell these stories.

So there's self and then there's culture. Two separate things. It's the self that wants to become perfect, and it's our culture that tells it what 'perfect' actually is. We'll soon discover, though, that these two forms are not quite as separate as they might first appear. For now, however, the journey of both self and culture must begin, and for that we have to travel back further than Aristotle and the early individualists, to a time before we were human.

BOOK ONE

The Tribal Self

Six foot four, shaved head, black T-shirt, goatee beard, tree-trunk neck, a heavy jewellery watch hanging loose around his wrist. *Huge*. John Pridmore sat square in front of me, his left fist planted knuckles-down on the arm of the chair. I'd arranged to meet him at his elderly mother's retirement flat, in Leyton, East London. We talked for hours, as the evening rose through the light around us and as his elderly mother, this delicate woman who'd once prayed for his death, listened in, occasionally interrupting with some remark or other.

I'd come to East London because I'd hoped this man's extraordinary story would shed light on some of the most ancient parts that make up the human self. Much of who we are today – what we feel, what we believe, how we think – can be traced back to an era before we were recognizably human. The first modern 'human' brain arrived in the fossil record around two hundred thousand years ago and yet for well over one and a half million years we existed as hunter-gatherers in tribes. It was during this time that our brain, and the self it produces, underwent much of its crucial development. We still carry this prehistoric tribal nature with us and John's spent much of his life unwittingly acting under its power. The tribish violence he took part in, as well as his preoccupation with status, hierarchy and reputation, welled up from some of the self's deepest cellars. As extreme as his experiences were, these basic instincts exist in us all.

His tale begins one evening when he was ten. He'd come home from Sea Scouts to hear his elder brother crying somewhere in the house. His dad was in the kitchen, looking angrier than he'd ever seen him. '*Get upstairs.*' He found his brother in the master bedroom, bereft and sitting on his parents' bed.

John's memories of before that night are of feeling loved: happy holidays at the beach in Hastings; long drinks of hot Horlicks; watching John Wayne films with his dad at the Granada Cinema in Walthamstow. His mother worked at the greengrocer's and his father was a policeman and would tell amazing stories about the ingenious crooks and celebrity villains he'd come across over the years: there was the burglar who'd steal cash from houses then drop it in the post on the way home to avoid getting caught. There was Roy 'Pretty Boy' Shaw, who it took eight men to arrest. Then there were the legendary Krays, who he'd stopped one day for speeding. 'He seemed to have more admiration and respect for villains than he did for policemen,' said John. 'The stories he'd tell us, it was a kind of fantasy world. I idolized them because of what he'd said.'

That night, when his parents eventually came to find the boys, his dad came straight out with it: 'You two are going to have to choose who you want to live with. Me or Mum.'

John didn't understand. He thought they were playing.

'Why?' he said. 'Is it a game?'

'No, it's not a game,' said his dad.

'We're getting a divorce,' said his mum.

What was a divorce? What were they going *on* about?

'But I live with both of you,' he said. 'You're my mum and dad.'

'Well, you're just going to have to choose.'

He looked at them, from one to the other.

'But I can't.'

John's father moved in with another woman. His mother moved into a psychiatric hospital called Claybury. He used to visit her there. 'She would either be in a daze and not recognize me or she'd be very angry and say, "You're the Devil's son."

Occasionally she'd be my mum and hug me and tell me she loved me. I think that was worse.'

John broke out in a nervous rash that covered his body from head to foot. The doctor ordered him to see her no longer. Back at home, life had become hard. His dad wasn't the same since he'd met this new woman. 'You never knew what you were walking into,' he said. 'It was like a tinderbox in there.' John began overeating. He started answering his father back. He started smashing his toys up. He started robbing coins from his father's ashtray and using them to play the machines. He started stealing napkin holders from Debenhams and money from school. He tried to run away. His dad and step-mum told him that if he didn't stop they were going to put him into care. 'You'd never know it if you met me but I was angry,' said John. 'Seething.'

One day, John and his friends broke into a pet shop to steal some white mice. The police came. When John went to court, he admitted to sixty counts of theft. He was sentenced to three months at the Kidlington Detention Centre in Oxfordshire. It was worse, in there, than he could ever have imagined. He learned to fight and he watched others fighting. He saw a visiting doctor beat up a boy because he kept wetting his bed. When he was released, he moved into a flat with his brother.

His mum, now recovered, had fallen in love with a man named Alan. John took a job at an electrical shop in Hoxton called Radio Unlimited, but he felt no duty to his new bosses. 'I felt as if anyone you trust, anyone you love, they just betray you.' He began stealing from the till, and from elsewhere, and was sentenced to three months at Hollesley Bay Young Offenders Institution in Suffolk. He was nineteen. When he arrived, a boy named Adrian demanded a share of his wages. A load of other lads were there, watching. John knew, from his time in Kidlington, that this was a moment that would make his reputation and nothing, in that place, was more important than that. He listened to Adrian going on and on, about all the ways in which he was supposed to suck up to him. 'He went on for five or ten minutes and I just smacked him. You either fought these guys or you ended up paying them everything.'

They put John in solitary confinement. Twenty-three hours a day, a bed, a toilet, a washbasin, and John observing his own mind breaking. 'Everything you do in life is to distract you from yourself,' he said. 'Suddenly you're just faced with you.' From his cell window, he had a distant view of the North Sea. He wrote letters to his mum and then his dad, telling them his 'life was a failure'. He hated himself. He'd spend hours watching the water, and people walking along the beach, and fantasize about killing himself. He wrote again to his mum, saying sorry for letting her down.

'I did think he'd let me down,' nodded his mother.

'Well, you didn't come visit, did you?' said John, not directly looking at her. 'I was very annoyed about that.'

'I would've gone because it's not so far, really, by train. But I think it was Alan that put me off. He never really wanted to do things for the whole day, did he? He used to say, "Oh, it's such a long way."'

I asked John if it hurt, listening to all this.

'No,' he said.

His mother, however, was beginning to look teary.

'It hurts me because I feel bad for John,' she said. 'I feel I've let him down, really.'

When he got out of prison, John met a man called Buller, who worked out of a second-hand office-furniture shop in Boundary Road, Walthamstow. Buller and his son also ran a business providing security for nightclubs and concerts. John enjoyed bouncing. He enjoyed fighting. But Buller had other interests too. One day, he asked John to pick up a Land Rover from Dover and drive it back to London. They paid him £5,000. He didn't know what was in it: drugs or guns or gold or whatever. But he did it, no problem, and the jobs rapidly became bigger.

John began to feel truly part of Buller's firm when he asked him to be present at a pub, for backup, during a meeting with a South London face that had the potential to end badly. 'You'll need to wear a black suit and black tie,' he was told. When John turned up, he couldn't believe it. There were at least sixty other men waiting there all dressed in exactly the same way. It was a display of power, of primal tribal force. It was a signal about his boss's reputation. When his adversary arrived, he walked through the door with six blokes. John grinned at the memory. 'I can still remember the look on his face.'

The firm that John was now a member of controlled the nightlife drugs trade in the West End of London. It did this by compelling club owners to use its security operation, the members of which would ensure only approved dealers were permitted inside the clubs, with others meeting intimidation and violence. John became one of their best enforcers. He wore a bespoke leather coat with special pockets sewn into the lining, one for his machete, the other for his CS gas. They complemented his portable armoury of a stiletto knife, a knuckleduster and a Jiffy lemon-juice bottle filled with ammonia. Every day was violent. Debts had to be honoured and rival gangs kept at bay. But it wasn't just groups of competing criminals that John and his associates would look down upon. 'To us, everyone in the general public was just complete idiots,' he said. 'Why would you go to a nine-to-five job when you could earn so much money doing so little? Our mentality was that we were the only sensible people and everyone else was vermin.'

*

If you'd have asked John, back then, why he was doing as he was doing, he'd have given you answers that might be tough to empathize with, but would, at least, be essentially rational: he was trying to raise his status within the firm because that way he'd be rewarded with money and women; he was beating this man up because he was a drug dealer who shouldn't have been operating on his turf; his gang, and their way of life, was good because they made a bounteous living without having to work too hard. He'd be unlikely to tell you, 'It's because I'm compelled by primal mechanisms within my self.' And yet that would, in a way, be the truest answer of all.

The essential rhythms of John's new life, of hierarchy, territory, tribal politics and bloody struggle for status and resources, are among the most elemental within the human self. Humans have spent more than 90 per cent of their time on earth existing in groups as hunter-gatherers and these basic instincts live on in all of us. If we're to understand who we are today, we should start by getting a glimpse of who we were back then. One way we can do that is by comparing our behaviour to that of the chimpanzees. We share a common ancestor, and more than 98 per cent of our DNA, with these animals. Along with the bonobo, they're our closest living relatives. By observing behaviours the human self shares with the chimp self, we might find clues as to which parts of us are so old they predate our ascent to the top of the world.

And you don't have to look for long until you start to recognize our ape parts. We two creatures turn out to have many eerie similarities. Like us, chimps are political animals. They exist in troops that are similar to, if about two-thirds smaller than, the tribes we spent many hundreds of thousands of years living in. This means that much of their daily life involves attempts at controlling their fates by manipulating those around them. They hide their feelings to get their own way. They hold long-term grudges. They negotiate peace by bringing aggressors together. They have a sense of fair play, protesting if given a smaller share of food than their neighbour as a reward, and are motivated to punish the selfish.

But it's in the chimps' preoccupation with hierarchy that we really begin to see the kinds of behaviours that are all too typical in human groups, such as John's. Weaker and younger chimpanzees regularly engage in conspiracies and coups: groups of low-rank individuals work in cooperation to foment dramatic and dangerous attempts at toppling the leadership. They keep track of political alliances: if one chimp defends another, it will expect that chimp's support during later conflicts. Failure to maintain this code of honour can bring on a crisis that leads to the breakdown of their coalition. They engage in political beatings and murders. These acts of violence are not the product of a moment of animalistic rage but are deliberately planned and carried out.

When the chimp has finally fought his way to the top of the tribe, he doesn't lead purely through violence. He must also be a canny politician. Alpha chimps studied by the famous primatologist Professor Frans de Waal have been seen switching tactics, once they'd achieved their place on the throne, from a focus on winning fights and cultivating relationships with other strong males to supporting the weaker members of the tribe, preventing them from being attacked. 'Chimpanzees are so clever about banding together that a leader needs allies to fortify his position as well as the greater community's acceptance,' he writes. 'Staying on top is a balancing act between forcefully asserting dominance, keeping supporters happy, and avoiding mass revolt. If this sounds familiar, it's because human politics works exactly the same.'

De Waal's account of the behaviour of a 'magnificent' alpha male called Luit gives us a tantalizing look at the qualities our closest cousins see as part of the 'ideal' self. 'Luit was popular with females, a mighty arbiter of disputes, protector of the downtrodden, and effective at disrupting bonding among rivals in the divide-and-rule tactic typical of both chimp and man.' This particular combination of strength, wisdom and caring is, of course, instantly recognizable as that which humans still exalt.

So chimps are like us in that our models of the ideal self are very similar, at least in outline. What we also share is our preoccupation with hierarchy. We have this preoccupation because our tribes, like theirs, have hierarchies that are *fluid*. An alpha male's reign usually lasts less than five years. This means we're constantly surrounded by intrigue and rumour. There are plots and victories; blood and drama. We're intensely interested in status, partly because that status has a high capacity to change. What our species also have in common is that members of our tribe band together to attack different tribes. The biological anthropologist Professor Richard Wrangham has observed that chimps and humans share 'a uniquely violent pattern of lethal intergroup aggression . . . Out of four thousand mammals and ten million or more animal species, this suite of behaviours is known only among chimpanzees and humans.'

So we're tribal. We're preoccupied with status and hierarchy; we're biased towards our own in-groups and prejudiced against others. It's automatic. It's how we think. It's who we are. To live a human life is to live groupishly. Laboratory experiments have found that humans, upon meeting someone new, will automatically encode just three points of information about them. What are these three things the brain considers so fundamentally vital? They are age and gender, which are essential for basic social interaction, and also race, which isn't. It's been found that babies universally prefer faces of their own race. Children as young as six, when shown pictures of people from other races in ambiguous situations, will tend to assume they're up to no good. Our era as hunter-gatherers might have begun to end around twelve thousand years ago, but our brains still live in this mode. Despite our knowledge of the hells this mode of thinking unleashes, we are helplessly groupish, ruthlessly dividing the world between in-groups and out-groups. We just can't help it.

The effect of our tribal brains has been shown in numerous experiments by social psychologists, who've found that all you have to do to generate spontaneous prejudice and bias in humans is to randomly divide one group of them into two. I've experienced the effects of the tribal self many times in my own years of reporting, everywhere from South Sudan where, caught up in the tribal civil war, I was abducted and just about escaped being shot, to the narco-zones on the hills outside Guatemala City. There, I visited a barrio called Peronia with a young man called Rigo Garcia. Peronia is a 'red zone': a concentration of maximum danger within a city of maximum danger that is, in turn, the capital of an outstandingly violent country. (At the time, the murder rate was more than double that of Mexico.) Rigo told me about his schooldays and the boxing club that opened up nearby. It was a fun and relaxed place. Boys and girls would come and learn the sport, some in the morning and some in the afternoon. A friendly rivalry developed between the two groups. Gradually, it became less friendly. Boasts became threats. People brought bats and machetes to class to defend themselves. One day, the afternoon group stormed the morning group's lesson. The terrified teachers locked the morning group inside. The kids began to build '*hechiza*': homemade guns, constructed from plumbing pipes and tubes from metal television stands. The morning group and the afternoon group had become gangs. 'Practically all the boys I went to school with have been killed,' Rigo told me, with a chilling nonchalance. 'One of them had his head cut off.'

We often hear this kind of behaviour called 'mindless' but there's a sense in which it's anything but: tribal aggression is an utterly predictable product of the human self. It's how it's built. It's what it does. As I listened to John tell his story, it was impossible not to be reminded that, despite the din and wizardry of modern life, despite how separate we feel from the beasts, the truth is that we are great apes that sit in the primate superfamily Hominoidea. We are modern yet ancient, advanced yet primitive. We are animals.

*

In John's job, as an enforcer, he had to control people. One of the other things John had to control was his conscience. He controlled it, mostly, by partying. Sex and cocaine made for lovely painkillers, as did flashy cars. He drove a 7 series BMW and a classic white Mercedes and had a penthouse flat in St John's Wood that overlooked Lord's Cricket Ground. He went to champagne and crack parties in Notting Hill. He'd get free drinks and peer respect and women's phone numbers everywhere he went. You didn't have to do anything, when you'd earned the kind of name for yourself that he had: the girls just came at you. He slept with so many, he couldn't count them. Life was amazing. He'd become a top villain, just like those characters his dad used to tell him about, when we was a boy. He was raising his status, fighting his way up the hierarchy, getting closer and closer to the top of the tribe.

One night, he was posted to work the door at the Borderline club on the eastern edge of Soho. It was a small place, frequented by the famous, and sometimes hosted secret gigs by popular international acts such as REM. John had a crush on the guest-list girl. It was a relatively common thing for people to claim to be on the guest list, only to find their names absent. The kind of place it was, there wasn't ever any trouble. Until one night, when there was.

'You're not on the list,' said the girl.

'Yeah, well, we're coming in anyway,' said the guy.

'You can't come in unless you're on the list,' said the girl.

John looked at the two men. They were no physical threat to him whatsoever. One of them noticed his gaze. 'And you're not going to stop us.'

'What they were doing, in my eyes, was belittling me in front of this girl,' said John. 'She's stunning, so I'm trying to get off with her. And the last thing . . .' He shook his head. 'I've only got my reputation, and she's seeing these two muppets aren't respecting me. They're taking it away. So I get a bat from behind the bar and I beat them. And I can honestly say that I nearly killed them.'

The most important lesson Buller ever taught his protégé

was the importance of reputation. 'For us it was like, you're not really earning money, you're not really earning women, what you're building is your name. That's the system of hierarchy in that world. If you're not looked up to as being the most hard, the most strong, the most vicious, if you lose your name, then you're nothing. But it was very strange. I'd be indoors watching *Little House on the Prairie* and have tears rolling down my face, but then I'd go to work and beat someone senseless. It was like two different worlds. I remember one time, when I was first starting out, I was in the pub with Buller and there was this little guy who was always trying to get some work or something off him, and every time he used to see me he'd say, just jokingly, "Hello, Lanky." One night, I picked him up by the neck and said, "If you ever belittle me in front of anyone again, I'll tear your head off." And Buller said to me, "Now you're beginning to understand what this is all about."'

*

With his obsessive interest in reputation, John was once again being silently manipulated by his tribal self. But in this, he's hardly alone. Caring about what others think of us is thought to be one of humanity's strongest preoccupations. Children start attempting to manage their reputations at around the age of five. Of course, in our hunter-gatherer days, it was critically important that we maintained a good reputation. Those who earned bad ones could easily be beaten, killed or ostracized – which would, in that environment, have been a likely death sentence. And still today, a core activity of the human self is maintaining a deep interest in, and trying to control, what others think of us. We're all, to some degree, anxious and hyperactive PR agents for our selves. When we realize our reputation is bad, the self can enter a state of pain, anger and despair. It might even start rejecting itself.

Reputation lives in gossip. It's in the delicious little tales we tell each other that our reputations – these radically simplified avatars that represent us in the social world – are given life. Our appearance as characters in moral stories means we're automatically cast as heroes or villains, our flaws or attributes magnified, depending on our role in the plot. And we can't help but gossip. Studies that measure how much of human conversation it constitutes put the figure at between 65 and 90 per cent. At the age of three, children start communicating their ideas of who can and can't be trusted to those around them. Despite the gendered stereotype, men are no less gossipy than women – they just tend to do it less when women are around to hear them. A study of gossip at a Belfast school found a large majority concerned individuals who'd broken some moral code; praise was relatively rare. One team of researchers found negative gossip can even affect our vision, causing us to automatically attend more to its subject.

Our helplessly gossipy nature is another inheritance from our tribal pasts. The anthropologist Professor Robin Dunbar has famously attempted to calculate the size of the human tribe, as it would've been back then. 'Dunbar's number', as it's known, came out at just below 148. Imagine being born into a tribe of 148 people. How would you keep track of them all? How would you find out who were the good people and who were the bad; who were the ones who'd share the meat and who would steal from you and stab you in the throat? A little bit of tittle, a little bit of tattle, that's how.

But gossip wasn't just a way of gathering crucial intelligence. It was also there to police the tribe. Gossip about a person breaking important rules would've generated powerful feelings of moral outrage in its members which would, in turn, have been likely to lead to severe punishments. This pattern, of course, is still all too evident in today's age of perfectionism, in which gossip about others, notably on social and online media, spreads quickly, leads to upswells of highly emotional moral outrage which, in turn, leads to shaming and calls for harsh punishment and then broken careers and broken people.

As virtuous as members of the mob sincerely believe themselves to be, when they're swept into this behaviour, they're actually being driven by brutal and primitive forces. They imagine themselves as angels when they are, in fact, as apes.

All of which brings us to a crucial place in our journey. It's in these ancient tribes that we begin to sense the deepest causes of our modern feelings of perfectionism. Because we didn't just crave a good reputation in order to avoid being beaten up and abandoned. We were – and still are – far more ambitious than that. We also wanted to be perceived well by others so we could rise up the hierarchy of the tribe. Our chief concerns, in the well-known words of the psychologist Professor Robert Hogan, were 'getting along and getting ahead'. We wanted to *get along* with others, by making a good reputation, and then use that good reputation to *get ahead*.

But how did we know how to get that good reputation in the first place? How did we learn what qualities our tribe valorized, and what it hated? We'd do it, in part, by listening to tribal gossip. It was in these moral outrage-making tales that we'd find out who we had to be if we wanted to be successful. So here we have it: ambitious selves, on the one hand, wanting to become perfect and, on the other, a kind of cultural group-concept of the 'ideal self'. These are the two separate forms we've been chasing.

It is possible to uncover the outlines of who the ideal self was, back then, in that distant epoch – what the basic characteristics were of a good member of the tribe and a bad one. We can do this, as crazy as this might sound, by experimenting on babies. The idea is that any functions and tendencies that are already in place, when we're born, must be absolutely fundamental to the self and, therefore, have an extremely deep history. These are the aspects, as child psychologist Professor Paul Bloom explains, that 'are not acquired through learning. They do not come from the mother's knee, or from school or church; they are instead the products of biological evolution.'

With these child experiments, researchers believe they've revealed our basic rules concerning what a 'good' person and a 'bad' person actually is. One series of tests with pre-verbal children involved a simple puppet show: a ball is trying to climb a hill, a kindly square is behind the ball, pushing it up, whilst a nasty triangle attempts to thwart its mission by shoving it back down the hill. Have six- to-ten-month-olds watch this show, and then offer them the shapes to play with afterwards and almost all of them will reach for the selflessly helpful square. These, writes Bloom, were 'bona fide social judgements on the part of the babies'.

A large body of such work now suggests that when humans say 'good', what they really mean is 'selfless'. We celebrate and reward those who are willing to make some sort of sacrifice for the benefit of others. You can see why this might've made sense from the perspective of the tribe, when sharing resources, including food, knowledge, time and care, would have been essential. The underside of that, of course, is selfishness, a trait which would have been energetically, and sometimes violently, discouraged.

Of course, this broad outline of our perfect self hasn't changed. We still like selfless people. We valorize them, in our gossip and our wider culture. Child experiments have shown that toddlers naturally expect sharing between members of their group. They even keep track of the politics of sharing, and know when another person is indebted to them. The urge for the policing of fairness has been found in four-year-olds. When offered a smaller portion of sugary treats than their neighbour, they usually preferred that nobody got any sweets at all than accept the bad deal. Even at that age, we're prepared to suffer, just to see that others are punished for being unfair. (But we're hypocritical about it: those children were more likely to *accept* the deal if it favoured them.) What we're doing is trying to control the selfishness of others, thereby maintaining the smooth running of the tribe. And still, today, these deep tribal rules hold enormous power over who we are, and what we want others to think about us.

But hearing about all this, I was puzzled. What about John? He was rising up his tribe – getting along with his associates and trying to get ahead, just as humans have always done – but he could hardly be described as 'selfless'. Didn't John's story spectacularly contradict what these social scientists had been arguing?

The answer didn't become fully clear until I understood the extent to which the behaviours we class as 'selfish' or 'selfless' are mediated by our tribal brains. Remember those toddlers who just naturally expected members of their group to share with each other? They *weren't* surprised that a person would refuse to share with someone from a different group. Selfless acts are most often made *on behalf of our people*. From John's point of view, he felt he was selflessly putting himself at constant risk of violent attack and imprisonment, in order that he could serve his gang better. From their perspective, he was being selfless. He was striving to be a person who was of *most benefit to his tribe*. The celebrated mythologist Joseph Campbell explained this principle well: 'Whether you call someone a hero or a monster is all relative to where the focus of your consciousness may be. A German [fighting in WWII] is as much a hero as the American who was sent over there to kill him.'

So here we are, back to our tribe and our tribishness. When we feel as if we want to become perfect, it's largely our tribe that defines, for us, what 'perfect' actually is. One of the ways it communicates this definition is via gossip, which often focuses on people breaking some important rule. Because of our tribal roots, all humans share the basic principle that a good person is selfless.

These are the universal basics. Today we might not live in literal tribes, but we do exist in psychological ones. We're all members of many overlapping 'in-groups'. We might identify as black or Asian, for example, or as a boomer or a millennial, or as rural or urban, or as iOS or Android. And we no longer have to rely solely on gossip to show us what kind of person we ought to be if we want to get along and get ahead. We're immersed in a rich culture, in which these lessons are communicated in newspapers, films, books and online. Often these stories lead to strikingly similar outcomes to those ancient, dangerous tales: heroes (even actors who *play* heroes) are hailed and grow in status, whilst those that transgress our rules are punished, perhaps by violence or ostracism. Most of us would like to be thought of as a kind of hero – which is to say, we hope to have a good 'reputation' in the daily, fluid social stories of our tribe.

But there's one final, crucial point to make about reputation. Humans are self-conscious creatures. We're constantly watching ourselves, judging ourselves, just as other people are judging us. When we catch ourselves behaving in obviously selfish ways, our minds alert us with a sense of alarm we call 'guilt'. We begin to experience guilt before the age of one. It's not pleasant; we like to believe that we're good people, the ideal self that deserves to rise to the top of the tribe. As anyone who's suffered from painful perfectionistic thinking can testify, we don't only crave a good reputation amongst others, we also crave it with *ourselves.*

*

He reminded John of a zip. Murphy: Irish geezer. Limp, greasy hair. All mouth. Desperate to be a face, but he was just a little car thief, a little street dealer. *All fucking rabbit.* John, as everyone in the Oliver Twist pub in Leytonstone that day knew, was widely regarded as one of the most dangerous criminals in London. He'd just bought some cigarettes from the machine when he accidentally brushed past Murphy. Up at the bar, John popped the lid of the packet and chatted with the landlord about what was happening on the news. It was 1991, the end of the first Gulf War. Murphy came up behind him.

'When you bump into people, you're supposed to say sorry.'

John turned.

Fucking rabbit man.

'What did you say?'

Fucking zip.

'What are you?' said Murphy. 'Deaf as well as ignorant?'

John grabbed Murphy's throat. He threw him on the floor and slammed his fist into his head again and again and again. There was Murphy struggling on the ground. And there was blood, heavy in John's shirt. Murphy had stabbed him. Then, down his back, a kind of tingling. John looked up. Murphy's mate was standing over him with a Stanley knife. He'd slashed him.

The landlord's wife strapped John's wounds as the two men fled.

'You've got to get to hospital,' she told him.

'No way.'

He called his associate, Phil. 'Come straight away. Bring a gun.' Phil arrived with a .38. They drove to Murphy's flat, kicked in the door, and discovered his wife and three children watching television. John aimed his weapon at the woman's face. She begged them, 'I haven't seen him.' They waited, outside the flat, for three hours. No Murphy. They looked for him the next day, in the Beaumont where he sometimes worked. No Murphy. They asked around the local drug dealers. No Murphy.

Then, finally, nearly a year later, a tip-off. Murphy sometimes picked his son up from school. John waited outside for days. Still no fucking zip man. And then, he appeared, with his six-year-old boy. 'Murphy!' cried John. 'Remember me?' John levelled the Irishman with one punch. He knelt on his throat and began smashing at his face with his fists. As parents and children screamed around him, he picked up Murphy's head by the ears and began slamming the back of his skull into the pavement.

'You're murdering him!' someone said.

John dropped him. 'The next time I see you, I'm going to kill you.'

Days later, John was celebrating a large drug deal in the Beaumont Arms on the Catworth Street Estate when Murphy's father approached. He was in his sixties, and upset. 'You've traumatized my grandson, watching his dad get beaten up in front of him like that.' John picked up a pint glass and smashed it into the old man's face. Murphy's brother lurched in to defend his dad. John slashed him with a stiletto knife and glassed him. He glowered at the silenced drinkers. 'Come on then!' A fat man moved towards him. 'I don't care who you are, you can't be glassing a sixty-year-old.' But John was outnumbered. He left the pub, changed out of his suit, and phoned two associates. They returned with golf clubs. John beat the fat man senseless, leaving him sprawled on a pool table, before smashing everything he could see and threatening to kill the landlord if the police were called.

A few weeks later, outside the Nightingales club in the West End, John smacked a drinker with a knuckleduster and then watched his head explode with blood on the pavement. 'You probably killed the bloke, John,' tutted Buller, as he drove him home. 'You've got to calm down.' John sat alone in his flat, in the Beaumont Estate, Leyton, with a spliff and a can of Special Brew. There were swords on the wall, pizza boxes and pornography on the floor. The room was painted black. Buller was right, he *had* been a bit touchy lately. He needed to calm down; if that man really was dead, he'd be looking at ten years for manslaughter. What was required, he decided, was a week off.

As he thought this, in Capworth Street a few miles away his mother was saying a nine-day prayer to St Jude, the patron saint of hopeless causes, begging him to take her son. 'I was asking him to intercede to God for me,' she said. 'I just said, "I've been praying all his life and he's not changing. You can have him because I've had enough. He's evil."'

At around 9 p.m. John heard a voice. It was listing all the

bad actions he had ever been responsible for. The violence, the women, the drugs and the betrayals. 'This is your life,' it said. 'This is what you've done.' He thought it was the television. But how did it know?

He turned off the television.

The voice remained, listing all his sins, 'one after the other, coming at me, to the point where I felt totally and utterly condemned.' He knew it, then: what the voice was and what it was telling him.

He was going to hell.

John ran into the streets outside his flat. He was on his knees, saying the first prayer of his life: 'Help me!' He felt bathed in beatific wonder; in shimmering golden awe. 'It was the greatest buzz I'd ever felt,' he said.

'What, even better than crack?' I asked.

'Not even in the same ballpark.'

It was the early hours when he arrived on his mother's doorstep. 'Mum,' he said. 'Something's happened.'

'What?'

'I've found God.'

She looked at him, astonished.

'Found God?' she said. 'At one o'clock in the morning?'

John's mum said he could stay at her flat. Her partner Alan gave him a Bible. John lay down with it and read the story of the Prodigal Son. That son, who had strayed, sinned and returned . . . it was him. He wept. The night became filled with supernatural portent. Hell noises surrounded him: bangs and crashes and howling. 'It was very weird,' he told me. When the light finally came again, Alan, who'd also heard the racket, told him, 'The Devil was very angry about you last night.'

John asked him if there was a place he could seek confession in total anonymity.

'Westminster Cathedral,' said Alan.

He took the Tube, queued behind a nun, and once he was

safe in the box of shadows, he began, listing the worst things he could think of.

'What prayers do you know?' asked the priest when it was all over.

'I know the Our Father.'

'Well, say an Our Father then,' the priest said. 'Welcome home.'

As he walked out of the cathedral, he felt as if he wanted to dance. Weeks later, he said a fuller confession to a priest at Aylesford Priory in Kent. It lasted for hours. He took more and more confessions, at one point clocking up four in a single day. He started making penances: deliberately depriving himself of sleep with Bible reading; going without food for days at a time; walking seven miles to church through the streets of the East End, in bare feet. But for all the fury and magic of his conversion, there were still things that confused him about this new world he'd found himself in. Like, how come the Roman Catholic Church was so rich, when so many were so poor? And how come the Pope behaved like the leader of some powerful tribe? Like a king?

John tried to put these quibbles to the back of his mind. He had much work to do on his self. He had to change. He had to become a better person. He realized, now, just how selfish he'd been. He prayed to God, 'All I've done is take. Now I want to give.'

*

It's often said that the self is a 'story'. If this is so, then, on the night of the Devil, John's story underwent an astonishing rewrite, turning him from violent gangster to worshipful Catholic. The events of that strange night hold essential clues for our journey – clues that will help us discover not only how the self is structured, but what kinds of events and patterns can lead to potentially catastrophic collapse.

To begin to understand what happened to John, we need to

consider just one facet of what psychologists and neuroscientists mean when they talk about this idea of the self as a 'story'. Doing so reveals something important and disturbing about the human self: that it is built to tell us a story of who we are, and that that story is a lie.

Consider, for a moment, what it is to be a conscious human. There are, essentially, four parts to the experience. Firstly, you have the experience of your senses – the sights, the sounds, the smells, the tastes, the physical sensations felt by the skin. Secondly, you have your sense of hallucinatory travel – your mind can summon images of past, future and fantasy. Thirdly, there is your emotional experience – that constantly churning ocean of fear, excitement, love, desire, hate and so on that writhes and swells beneath your days. Finally, you have your internal monologue, the chatty voice that narrates it all, interpreting everything that's happening to you, discussing it with you, making theories about it, never shutting up.

Now remember John's night of the Devil. What were the different components of his experience? First, he was hearing a disembodied voice. Second, he was remembering ideas about God, the Devil, damnation and forgiveness that he'd come across in his Western Christian culture. Third, he was feeling dread. Finally, and most importantly, he was hearing his internal monologue. His inner voice was tying all the disparate elements of his experience together and turning them into a useful story that would make sense of it all. It was saying, 'That horrible voice you're hearing is Satan. It means you're going to hell. But don't be scared, John, you know what to do. You need to pray to God for forgiveness.' John believes what happened to him, that night, was a visitation from the Devil. But for me, it seems more likely that he had a brief psychotic episode, and what saved him from having a full and sustained breakdown was that voice in his head, maintaining his sense of control, telling him what was happening and what to do next. His inner voice pirouetted across the madness, tying it all together, making it safe. Neuroscientists have a name for this voice – they sometimes call it the 'left-brain interpreter'. If the self is a story, then prepare to meet its slippery author.

The bizarre work of the self's interpreter was first properly exposed in the 1960s by a team that included cognitive neuroscientist Professor Michael Gazzaniga. They dreamed up an ingenious method of fooling it. They did this by taking advantage of epilepsy patients who'd undergone 'split-brain' surgery, in which the wiring that connected the brain's two hemispheres had been cut to prevent dangerous 'grand mal' seizures spreading across them. Amazingly, these procedures worked, and the patients were able to lead relatively ordinary lives. But because their brains had been split, and most of the word- and speech-making circuitry that the interpreter relies upon is in the brain's *left* half, the researchers realized they could implant ideas into the patient's right half without their inner voice detecting them. And if their inner voice didn't know they were there, it couldn't 'tell' its owner about them.

This is how it worked: they would angle pictures and videos in a specific way towards the patient's left eye and then, because of the way in which the brain's wired up, the images of those pictures and videos would travel into the patient's right hemisphere. But because their right hemisphere had no inner voice it couldn't say, for example, 'It's a photo of a chicken,' and so the patient would have *no conscious idea* that they'd seen anything, let alone a photo of a chicken. If a man's right hemisphere was shown a picture of a hat, say, he would deny having seen anything at all – but then be alarmed when his left hand (which, of course, is controlled by his right hemisphere), suddenly began pointing at a hat, apparently of its own volition. This actually happened.

One test involved a woman's right hemisphere being shown a frightening film of a man being pushed into a fire. All she was aware of, apart from a vague experience of some kind of flash, was that she suddenly felt emotions of fear. 'I don't really know why, but I'm kind of scared,' she told the scientists. 'I feel jumpy. I think maybe I don't like this room. Or maybe it's you. You're making me nervous.' She turned to an assistant. 'I know I like Dr Gazzaniga,' she said, 'but, right now, I'm scared of him.' Her left-brain interpreter, unaware that the film had triggered her fear, searched its environment for the first explanation that made sense of what it was feeling and produced that explanation as a fact: 'It's Gazzaniga,' it said. 'He's freaking you out.' The patient accepted the untruth as fact. 'The left-brain interpreter is driven to seek explanations or causes for events,' wrote the scientist. 'The first makes-sense explanation will do.' Along with his team, who called these stories 'confabulations', he demonstrated it happening again and again. When one man's silent right hemisphere was commanded, by a sign, to 'walk', he obediently stood up and walked towards the kitchen. When asked why he was doing this, he confabulated: 'Because I'm thirsty.' A woman's silent right hemisphere was shown a picture of a pin-up girl. When it was enquired of her why she was giggling, she too confabulated, telling the researchers they had a 'funny machine'.

Remember, these people weren't making these errors because they were 'mad' in any sense. All the surgery did was enable Gazzaniga to expose the work of the interpreter. The discomforting truth is that we all have interpreters narrating our lives, and they're all just guessing. We all confabulate, all the time. We're moving around the world, doing things and feeling things and saying things, for myriad unconscious reasons, whilst a specific part of our brain constantly strives to create a makes-sense narrative of what we're up to and why. But the voice *has no direct access* to the real reasons we do anything. It doesn't really know why we feel the things we feel and why we do the things we do. It's making it up.

Since these experiments, study after study has demonstrated ordinary, everyday confabulation in non-split-brain participants. My favourite involves people who were shown two photos of members of the opposite sex and asked to choose which they were more attracted to. The photos were then placed face down and, with a magician's sleight of hand, switched. The photos were then shown again, now in the wrong positions. Incredibly, only 17 per cent of participants noticed that they'd been switched. Those that didn't notice were asked to give a detailed explanation as to why they'd found *this* person more attractive, as opposed to the other one. They merrily confabulated away, listing all of the reasons, even though they were actually describing the wrong person.

Our brains invent stories like this because they want to make us feel we're in control of our thoughts, feelings and behaviour. The guesses the interpreter makes might be right. But they might just as easily be wrong. 'When we set out to explain our actions, they are all post-hoc explanations using post-hoc observations with no access to nonconscious processing,' writes Gazzaniga. Any inconvenient facts that don't fit with the interpreter's story are ignored or suppressed. 'That "you" that you're so proud of is a story woven together by your interpreter module to account for as much of your behaviour as it can incorporate, and it denies and rationalizes the rest.' All this means that 'listening to people's explanations for their actions is interesting but often a waste of time.'

If this is correct, it leads us towards an uneasy conclusion. Imagine, for a moment, that a human works like this: you're a zombie, just behaving automatically, sometimes chaotically, and the only reason you think you're in control of your behaviour is that you have a dishonest voice in your head that tells you that you are. When you do something, like decide who you fancy, or beat someone nearly to death outside a London nightclub, your little voice tells you these acts were the result of conscious decisions made by you, and then lists all the reasons you were right – but actually you're like a zombie, with no free will, being fooled into believing you have the ability to make conscious choices by your lying little voice. Wouldn't that be odd?

Well, yes. It's also widely believed to be true. The disconcerting fact is that what we do in the world is, depending on which academic you believe, either largely or wholly controlled by our unconscious. 'If you devote your time to thinking about what the brain, hormones, genes, evolution, childhood, foetal environment and so on have to do with behaviour, as I do,' writes the professor of biology, neuroscience and neurosurgery Robert Sapolsky, 'it seems simply impossible to think there is free will.' Most of those who argue that we *do* have free will believe its power to be limited, marginal or conditional. The illusion that we have it in the way we think we do is perhaps the most important and most devious job of the self.

Confabulation, as it emerged on the night of the Devil, brings with it two ideas that are important for our journey. First, the extent to which the self is a 'story'. It turns the chaos of the outside and our inner worlds into a highly simplified narrative which, if we're mentally healthy, serves to reassure us that we have control and all is well. For the person struggling with perfectionist thinking, of course, that voice can sometimes be more enemy than friend: *you're feeling anxious and sad because you're not good enough, you're a failure, you're a prick, you're fat and ugly and you always will be.* These story-making processes are universal. All humans' brains are structured in this way, because that's how they've evolved.

But there's another clue, in John's experience, that leads us to the next stage in our journey. During the night of the Devil, John's mind grabbed the 'story' that would form the structure of his new life from his culture. He was raised in a Christian country, by a Catholic mother. His plan for the future and his replacement identity would be built from ideas from these sources. Tales from his culture were selected by his self, which rebuilt itself according to their design. This hints at the incredible power culture, and the particular stories it surrounds us with, can have over us. It also suggests that the separation between 'self' and 'culture' might not be so much of a separation after all.

*

'In life, if you're in control, you feel you can't be hurt,' John told me. 'If I felt I was out of control in a situation, the fear would make me very angry. In that anger I'd have control. People couldn't hurt me.' If John's found happiness now it is, at least in part, because he's subcontracted his need for control to God. 'One of the biggest changes in my life is that I don't feel fear much now,' he told me. 'The more you fill yourself with God, the less fear there is. And the less I'm in control and more I let God be in control, the more peaceful and patient I am.'

Before I left him and his mother, I wanted to get some idea of how much he'd actually changed. Had there been a true metamorphosis, or was it really just the story John's brain had created for him to live in that had changed? 'I'm not perfect,' he said. 'I can be incredibly selfish. I'm lustful. I still get wound up.' He thought for a moment. 'Definitely anger is one. If I feel like I can't get through to someone, that anger is me trying to control the situation so they get my point of view.' He thought again. 'And if I see someone unjustly treating someone else, or swearing in front of a woman or that kind of thing.' He looked at his mother. 'There was that incident about a year ago where that guy spat at you.'

'At least the car window was shut,' she said.

'What happened?' I asked.

'This guy spat at my mum whilst I was pulled up at some lights.'

'And what did you do?'

'I got out and clumped him,' he said. 'I sent him and his phone flying down the road.'

'But you didn't feel like you were out of control?'

'Not at all,' he said. 'Before I found Jesus, I would've really hurt him. I wouldn't have stopped.'

BOOK TWO

The Perfectible Self

It would usually begin with my walking past a parked car. I'd catch my reflection in its window and glimpse something horrible jutting out of my shirt. No, I'd reassure myself. It's just a distortion. It's the shape of cars, isn't it? They bulge in the middle. It'd be a van window, next, then an estate agents, then a large supermarket and then, if I wanted best-in-class accuracy, an Apple Store, my eyes surreptitiously darting towards the panes of glass, each time examining my shape with a growing sense of despair. By this stage, I'd have already been through weeks of denial: 'My stomach's swollen because I'm hungry'; 'My stomach's swollen because I've just eaten'; 'The tumble dryer's shrinking my trousers again'. I'd find myself crossing my arms in the company of slender colleagues; sucking myself in near cameras and young people. In the perilous naked moments, when undressing for bed, my eyes would fix themselves hard on the curtains. Soon afterwards, a new diet would start and I'd be miserable for three months. Then the whole cycle would begin again.

I don't diet so much any more. (Literally, as I'm writing this, I'm midway through my morning dose of 'Huel', an attempt at making a nutritionally perfect form of 'human fuel' – surely a marquee product of our age of perfectionism.) Although I'm attempting to cut down on the junk food that's one of my life's great joys, I've also been trying to accept the fact that I'm no longer twenty-three. What I tell myself is that I'm forty years old and I've earned the right to have a bit of a belly. I really honestly believe this. But what I *feel* is something completely different. All too often I'll find myself grabbing at the squinch of stomach meat that spills out over my belt and pulling at it. The lardy bib I keep beneath my shirt is not so much a body part as it is a psychological flaw that's become material – shame that I can touch. My weight tells a story of who I am: the person who likes to spend weekends sunk into the sofa, surrounded by pizza boxes, sticky-fingered and burping, a grown man who might as well be wearing a nappy. My fatness, the outline of my body, can make me feel as if I'm guilty of a moral transgression.

This bizarre notion – that physical appearance and moral worth are directly linked – is so deeply sunk into my brain that I feel it to be true on an emotional level. I feel it so powerfully that it easily overwhelms the part of me that reasons that it's nonsense. But it's a cultural invention. It was made up by humans. As classical scholar Dr Michael Squire of King's College, London told me, 'The Ancient Greeks had this idea that being physically beautiful was the same as being ethically good and, likewise, being physically ugly was the same as being ethically bad.' They had a word for this: *kalokagathia*, which came from *kalos*, meaning beautiful, *kai*, meaning 'and', and *agathos*, meaning 'good'. 'This idea, that the bodily form is inherently important for understanding who someone is, is very much still with us,' he said. The scholar Professor Werner Jaeger has written of *kalokagathia*'s roots in early Greek aristocracy, describing it as their 'ideal of human perfection, an ideal towards which the elite of the race was constantly trained'. Just as John Pridmore's self snatched core ideas of who he was *and who he ought to be* from his culture, so does mine.

It's odd, and hard to accept, that a great deal of who we so intimately feel we are is the product of the thoughts and experiences of long-dead people. It's like removing a chunk of your face and realizing it belongs to someone else. According to social scientist Professor John Hewitt, a part of the reason we find this difficult is that, over the last century or so, our definitions of what a person is and what causes its behaviour have come largely from psychology. 'The brains we have and the psychology those brains produce is important,' he told me, 'but it doesn't explain many of the things we'd like to explain, whereas culture and society do.' To what extent, then, are we the culture we're born into? 'In one sense, I'd say culture is about 90 per cent of what we are.'

The self's ingestion of culture can be tracked, in a startling form, in the brain of the growing baby. Despite the fact we're born with almost as many neurons as we'll ever need, the weight of a child's brain increases by more than 30 per cent during its first fifteen months. If this rapid gain isn't due to the generation of brain cells, then what is it? Most of it is the weight of new connections, or synapses, that are forming between those cells. By the age of two, a human will have generated over a hundred trillion synapses, double that of an adult. So great is this extra brain functionality that youngsters even develop cognitive powers the rest of us lack. Six-month-olds can recognize the faces of individuals from other races with an ease that would have the rest of us worrying quietly whether we were racist. They can even readily identify *monkey* faces. Babies can hear tones in foreign languages that their parents are deaf to. They're also thought to experience synaesthesia, the eerie blending of the senses that enables people to taste colours, and so on.

But then begins the cull. These connections start dying off at a rate of up to 100,000 per second. It's believed that this is one of the ways the brain shapes itself to its environment. Huge connectivity means it's prepared to deal with a wide range of potential possibilities. Then, when connections between neurons are not activated, they're killed. They call this 'neural pruning', and it works a little like a sculptor carving a face into a block of marble: it's what's taken away, not what's added, that really makes us who we are.

When we're born, then, our brain is ready for the world – or at least *a* world. It rushes out to greet it, gets to know it, then prunes itself down, specializing itself for the particular culture in which it finds itself. Much of the environment's influence over who we turn out to be takes place in childhood and adolescence, periods in which our brain is in its phase of heightened, developmental plasticity. Influencing the way our brains initially wire up are our genes. 'But a genome doesn't specify the final form of the brain,' Professor Jonathan Haidt told me. 'It really just specifies the starting conditions. That's sort of the initial direction, the first draft of mind. But then as we grow up our brains expect to get all kinds of information from the environment. They continue growing, while incorporating that information.'

People have long argued about what is more important in determining who we are, genes or environment. In a major study, researchers in Queensland collated the results of 2,748 papers and concluded the average variation across all human traits and diseases is caused by 49 per cent genetic factors and 51 per cent environmental factors. Co-author Beben Benyamin added that environment seemed to have a greater influence over aspects related to 'social values and attitudes'.

But what's also known, these days, is that nature and nurture are not at war, battling for control over human minds and bodies. 'When I was an undergraduate, over twenty-five years ago, you heard about genetics and environment being spoken about as two different things that were impacting on your development,' the neuroscientist Professor Sophie Scott told me. 'Now we've come to realize that it's much more complex than just being a dollop of genetics and a squirt of environment.' The relationship is symbiotic. Nature and nurture are not in competition, but in conspiracy. That's not to say, however, that great questions don't remain about exactly how the environment changes us. 'We're very good at modelling genetics because we have lots of good ideas about how genes work,' the psychologist Professor Chris McManus told me. 'But try and do the same thing for the model of the environment and we don't have one. We have very little information about how the environment actually affects things.'

Included in this broad term 'environment' is individual experience. The things we go through, not least the traumatic ones, clearly have an impact on how our lives turn out. But, as John Pridmore demonstrated with his theory about how his parents' divorce changed him, we're often well aware of these effects. Especially since the popularization of therapy, we're used to our left-brain interpreters using the things that have happened to us as plot points in the story of our lives.

Culture's influence on the self is more insidious. It comes at us from many directions: from our family, who'll share their values and beliefs with us as we grow up; from our friends and associates, not least during adolescence; and also from our 'social category' – our gender, class, race, and so on – whose cultural norms we'll be susceptible to absorbing. It's also delivered in the many vehicles we more traditionally think of when we talk of 'culture' – at church, in the cinema, on social media, on television, in books and newspapers. Culture can be seen as a web of instructions, like computer code, that surrounds and saturates us. It tells us what a person should be – what it looks like, how it behaves, what it wants. We internalize these rules, then begin adhering to them as if they were laws of the universe. When I feel an emotion of revulsion because my stomach is a long way from the 'ideal' shape, that's my culture talking. I've absorbed it. It's *inside me*. To a significant extent, it controls me, like a parasite, admonishing me when I stray too far from its models.

It's both remarkable, and rather depressing, that the body ideals of Ancient Greece look so similar to ours today. Indeed, 2,500 year old depictions of Hercules and Adonis could feature quite happily on the cover of next month's *Men's Health* magazine, even down to the pelvic V-line. Yet peer outside the bubble of the West and things can become disorientatingly different. Professor Sophie Scott told me about a friend who used to spend time in Tanzania, collecting data. 'Being fat there is a sign of status,' she said. 'People would comment in a negative way if she lost weight. Then, when she'd come back to the UK, everybody would go, "Oh my God, you look amazing! You've lost so much weight!" It's very, very hard to think yourself out of the stuff you've grown up in.'

But culture's reach descends far deeper than this, as Sophie discovered when she took her neuroscience lab to northern Namibia to meet the Himba people. 'They have a Stone Age lifestyle,' she said. 'They're not contaminated by our culture, which was why we wanted to work with them.' Sophie's team deliberately designed a very simple study. The Himba would hear two sounds then a third sound. They'd have to say which of the first two sounds was expressing the same emotion as the third. 'As soon as we got there, we had to completely redo all our tests,' she said. 'They just had no idea what we were talking about. That was when we started to realize, the vast majority of the experiments we do in the UK are with people who've been through formal education. Just holding information in working memory, then thinking about it, manipulating it, and making a response – that's something you learn to do. And you know what? There's loads of things they can do that we can't.' Children growing up in Tanzania, for instance, are able to navigate a landscape that looks, to the average Westerner, utterly featureless, and have measurably better spatial memory than we do. 'No one's taught them to do that. Their environment's meant they've had to learn.'

Our obsession with youth, too, turns out to have a cultural root. Sophie described a body of scientific work that shows that, when you ask people to talk about their lives, they don't tend to bring things up randomly from across their life. Instead they tell you mostly about events from their twenties. 'It's thought that your brain might work differently in your twenties and you're laying down more memories,' she said. 'That's part of the explanation. But more recently somebody did an experiment where, instead of asking older people, they spoke to people aged ten to eighteen. And it turns out, they do the same thing. They talk about what's going to happen in their twenties! Basically, in our culture, we think being in your twenties is brilliant.'

John Pridmore's experience showed us some of the features of the human self that we all share, no matter where we're from: our tendency for groupishness; our gossip–outrage–punishment pattern of social policing; our valorization of selflessness and hatred of its opposite; our restless desire to get along, which gives us prestige, and get ahead, which gives us status; our 'storytelling' brain with its confabulating narrator that, if it's working well, gives us a feeling of control over our inner and outer world. These are some of the self's most ancient parts. But, above the pistons and pipes of this basic human mechanism, there are layers and layers of intricately tooled machine-work – the richness and detail that build up to make us who we are. A large proportion of this machine-work is culture.

It's thought that we started to become recognizably cultural animals around 45,000 years ago. But if you, like John, were born in the West, many of your finer and most important cogs, springs and wheels were forged around 2,500 years ago, amid the spectacular drama and muscular beauty of Mediterranean mountains and sea.

*

The land of our cultural birth could be hell for those who relied on it. Aside from the remote northern plains, just a fifth of Ancient Greece was well suited for agriculture, with much of the rest jagged mountains, islands and inlets. By 500 BC, even the forests had been mostly cut down for timber. Irrigation was impossible; drought was a lingering, fatal menace. The soil was poor. The people survived, largely, on their own wits, or as part of tiny, near self-sufficient industries. Many managed by hunting, foraging, stock-raising or running their own small farms. Others produced olive oil, animal hides, chestnuts, pottery, and wine. But what enabled Ancient Greece to thrive, and what went on to sire its complex and advanced systems of class and wealth, was the Mediterranean. 'We dwell about the sea like ants or frogs around a pond,' as Socrates said. That 'pond' was their saviour and creator.

Their bad land pushed them out onto the waters. The Greeks were daring travellers – exporters and importers. They were pirates and they were hustlers. They traded with each other on coastal routes, and by heading onto the perilous oceans, to Egypt and the Near East and beyond, to make contacts and deals. Their ports welcomed visitors from distant lands bearing strange products, new technologies and fresh notions that challenged the shibboleths of the locals. This realm of proto-entrepreneurialism, travel, novelty and debate was to form the beginning of what, according historian Professor Werner Jaeger, 'appears to be the beginning of a new conception of the value of the individual, that each soul is in itself an end of infinite value'. This idea – of the individual as a node of value that had the potential to improve itself – birthed the modern Western civilization of freedom, celebrity, democracy and self-improvement we live in today.

In its form, Ancient Greece was not a nation as we'd recognize it now. The ants or frogs of those serrated shores combined to form a 'civilization of cities'. This was a pointillist realm composed, chiefly, of more than a thousand self-ruling city states. They ranged in size from near-insignificant hamlets to those fizzling precincts of legend: Corinth, Thebes, Athens and Sparta. In faraway lands kings and tyrants claimed their authority from gods and maintained their rule by blood and dread. Such attempts didn't last so long in Greece. As the King of Athens boasts in Euripides' 423 BC play *The Suppliants*, 'You started your speech with a false statement, stranger, in looking for a tyrant here. For the city is not ruled by one man, but is free.' It was in Athens that, for half a century, freedom became the basis of a new political system: democracy. With the creation of a political class came the creation of satire, such as that of the 'father of comedy' Aristophanes, whose *The Babylonians* was denounced by the politicians it mocked as slanderous.

Of course, this 'freedom' was mostly enjoyed by a subset of men. But partial as it was, it remains an astonishing achievement, an epochal breakthrough in the long story of the human. Athenians were free to travel great distances to enjoy plays or poetry. People could up and leave work to compete in the Olympics. A commoner could debate royalty without fear of torture or death. If a person clashed with their neighbours, or the laws of their place of birth, they could simply move to another city and start again. The Greeks were active and the forces by which they could change their lives and their world lay within them.

The ordinary Greek would seek to control the gods by the giving of gifts and honours. Any reward for doing so would not come in some paradisiacal afterlife, but now. The closest thing they had to heaven was Elysium, invitations to which were not extended on the basis of moral worth but of status. Their version of hell was Tartarus, a slightly obscure realm that was reserved for those who'd committed the most grotesque crimes. It was in Tartarus that Sisyphus was punished by being compelled to push a rock up a great hill only to have it roll back down and for him to start again, for eternity. It could hardly be more Greek, this ultimate nightmare of having all your exertions come to naught, this catastrophe of agency thwarted. Writes the psychologist Professor Richard Nisbett, who pioneered the study of what he calls 'the geography of thought', 'The Greeks, more than any other ancient peoples, and in fact more than most people on the planet today, had a remarkable sense of personal agency – the sense that they were in charge of their own lives and free to do as they chose. One definition of happiness for the Greeks was that it consisted of being able to exercise their powers in pursuit of excellence in a life free from constraints.'

One of the highest possible achievements was being part of the political class and thereby contributing to the community. The average citizen of Athens saw the pursuit of excellence as a way of becoming a better and more useful member of society. In order to achieve this, they arrived at the epoch-making conclusion that *reason* was a more powerful tool than superstition. In the sixth century BC, Thales foreshadowed science itself by asking what was the single thing everything could be reduced to. (Water, he decided.) A hundred years later, Socrates obsessed over the fundamental nature, not of the material world like Thales, but of abstract truths. 'What is courage?' he demanded of the minds that surrounded him, challenging them on each further argument as it came. 'What is beauty? What is happiness?'

One of these minds, Plato, became preoccupied with questions about the ideal city-state, and believed in the existence of a metaphysical realm of pure form. Aristotle, his pupil, rejected this, insisting that the only reality is that which we can sense. We live in a world of things, he thought, and each of those things has unique properties that can be defined and categorized and acts predictably according to certain laws: an apple that falls to the ground does so under the force of gravity, he said, just as it floats on the sea under that of levity. His view of reality, and of change, was deeply optimistic. Historian Adrienne Mayor writes that Aristotle believed that 'all things in nature moved towards achieving perfection of their potentials'.

One thing in the world that undergoes change is a human. A person, like an apple, is an object in isolation that possesses its own unique properties. But what kind of thing is it? A sort of 'political animal', thought Aristotle. And this, crucially, was an animal that could be improved. It's for this reason, argues

Jaeger, that 'the history of personality in Europe must start' with the Greeks.

And so, here it was, the age of perfectionism in its emerging form – a culture of veneration and pursuit of the perfect human self. In Greek life, the talents of remarkable people were fetishized. Sublime statues depicted ideal masculine and feminine forms. Men would compete in spear-throwing, chariot-racing and bull-leaping. Skills in debating, which could take place anywhere, from the marketplace to within the military, were highly regarded. The spirit of competition sweated from the very skin of the citizenry, each glancing jealously at the other's success – 'Potter resents potter and carpenter resents carpenter, and beggar is jealous of beggar and poet of poet,' wrote Hesiod. Everybody wanted the glory of being the best for the prizes of meat and money, of course, but even more, for the fame and the glory. For the victor not to be honoured by all was considered scandalous, the denial of public honour 'the greatest of human tragedies'. But from this self-regarding culture bubbled up a warning. It came, naturally enough, in the form of a story; that of a proud hunter who glimpsed his image in a pool and fell deeply in love, only to despair and eventually die of sadness when he realized the object of his desire was but a reflection. His name was Narcissus.

For academics such as Nisbett, it all began with the land. The ecology of Ancient Greece silently and powerfully moulded a new way of being human. Its dry crags and hills and inlets and islands and poor soil and dangerous weather forced into being an economy of small-businessmen who relied on themselves and those closest to them to survive. It also conspired to form its physical structure, its network of city-states. Its seaward trading posts brought in new ideas and encouraged debate. The individuals that grew up in this civilization of cities then vied for the power to rule them. A person's worth, and success in rising up in society, depended largely on their own talents and self-belief. Celebrities were hailed. Beautiful bodies venerated. A particulate landscape became a particulate nation became a particulate people with particulate minds. 'The story is ecology to economy to social practices to cognition,' Nisbett told me. Our Western self is the son of this atomized world.

What had been created in Ancient Greece was individualism. As you'd expect in such an intellectually dynamic place, this was a concept that had many critics. But it's one that still dominates our lives today. In fact, it's so easy to spot the foundations of our twenty-first-century age of perfectionism in this Greek notion that I want to focus on its evolution – from its inception under Aegean skies to its supercharged and ever more perfection-demanding 'neoliberal' manifestation that beguiles us so today. This will be the journey of an idea – of how we in the West, since the days of Aristotle, have tended to see ourselves as individuals rather than part of a connected whole. It will explore the nature of this individual form of self, track how it's changed, consider why it's changed, and examine some of its major consequences.

I'm going to follow this story through just one chain of people, using their lives as stepping stones through the vast universe of our history. This will, then, be a rather partial and simplified tale. Further, it will mean ignoring chapters of our shared past that might usually be considered crucial. But I believe real light can be shed on some of the sadnesses that currently ail us by looking at a handful of the lives and times that managed to change, in some essential way, our idea of what it is to be a freedom-fetishizing, I-focused, individualist person. As well as Ancient Greece, this will mean touching down in the eras of medieval Christianity, the Industrial Revolution, post-war America and Silicon Valley. Each of these periods added something new and unique to the model of the ideal self that taunts us today.

But before we press on, an important question must be addressed. How can it be that the beliefs and values of a distant people, 2,500 years ago, can play a part in forming who we are in the twenty-first century? There are, no doubt, many answers to this question. For now, though, our investigation into how culture enters and then changes us must return to the idea of the self as a 'storyteller'. In doing so, we'll realize just how porous the boundary really is between the stories that surround us and the story that *is* us.

*

In many ways, we can't help but experience our lives as story. And it's not only the work of the confabulating interpreter that's responsible for this. Because of the way our brains function, our sense of 'me' naturally runs in narrative mode: we feel as if we're the hero of the steadily unfolding plot of our lives, one that's complete with allies, villains, sudden reversals of fortune, and difficult quests for happiness and prizes. Our tribal brains cast haloes around our friends and plant horns on the heads of our enemies. Our 'episodic memory' means we experience our lives as a sequence of scenes – a simplistic chain of cause and effect. Our 'autobiographical memory' helps imbue these scenes with subtextual themes and moral lessons. We're constantly moving forward, pursuing our goals, on an active quest to make our lives, and perhaps the lives of others, somehow better. To have a self is to feel as if we are, in the words of neuroscientist Professor Chris Frith, the 'invisible actor at the centre of the world'.

And our biased brains ensure that the 'invisible actor' that is us seems like a good person – someone morally decent whose values and opinions are usually correct. Just as in Ancient Greece, we like to think our lives have a pattern that's heading towards the perfection of our potentials: although we suffer setbacks, we're getting better and better, closer and closer to perfect. The healthy, happy brain runs a gamut of sly tricks in order to help us feel this way. It ensures we're often over-generous with our estimation of ourselves, imagining we're better looking, kinder, wiser, more intelligent, have better judgement, are less prejudiced and more effective in our personal and working lives than is actually true. One recent study that examined these kind of biases found that 'virtually all individuals irrationally inflated their moral qualities'.

Work by psychologists including Professor Nicholas Epley has shown an especially invidious bias in which we tend to cast ourselves in heroic light, whilst throwing shade on those around us. It centres on the two different sets of motivations people might have for pursuing their work. There are the heroic 'intrinsic' motivations, such as pride, the joy of learning new things and the accomplishment of doing something worthwhile. Then there are the more suspect 'extrinsic' ones, such as pay, job security and fringe benefits. Every year, Epley tests his business students at the University of Chicago. And he always gets the same result – one that shows, Epley writes, 'a subtle dehumanisation of their classmates. My students think all of these incentives are important, of course, but they judge that the intrinsic motivators are significantly more important to them than they are to their fellow students. "I care about doing something worthwhile," their results say, "but others are mainly in it for the money."' Other studies show similar results.

So the brain is a storyteller and it's also a hero-maker – and the hero that it makes is you. But the hero it makes and the plot it shapes your life around are not created in a void. The brain is a plagiarist, stealing ideas from the stories that surround it, then incorporating them into its self. Like John Pridmore and the ancient biblical tales he adopted, we absorb the stories that flow around our culture and use them to make sense of our past, our future, and to help us figure out who we are and who we want to be. We use them to construct our 'narrative identity'.

It's thought that the stories our parents tell us, and their characteristic shape, begin to play a part in building our understanding of self and life no earlier than the age of two. Between five and seven, the content of those stories – including ideas about cultural roles, institutions and values – starts to merge with our sense of who we *are* and who we *should be* in society. We now have a model of our 'cultural self.' It's during adolescence, according to the psychologist Professor Dan McAdams, that we start to understand our lives as a 'grand narrative'. In order to help build this narrative, our memories of the past are shuffled and warped – edited as if by a canny screenwriter who's turning us into a sympathetic, heroic character. We also start imaging the future in such a way that it fits the story we're creating.

The storyteller inside us, then, is heavily influenced by the culture it's immersed in. From the fairy tales we hear as children, to films and works of literature, to the documentaries and news stories that narrativize the world more directly, to ancient parables in holy books, stories work as both entertainment and a kind of shopping mall of the self. 'Culture provides each person with an extensive menu of stories about how to live,' writes McAdams, 'and each of us chooses from the menu.' We build our sense of who we are by 'appropriating stories from culture'. Turning our lives into myth, he writes, 'is what adulthood is all about'. Our story gives our life meaning and purpose. It distracts us from the chaos and hopelessness and dread of the truth.

But in the basic shapes of modern storytelling it's also possible to burrow *beneath* culture and find basic plumbing that was laid millions of years ago. Joseph Campbell, the mythologist who's perhaps had more influence over popular Western storytellers than anyone in the last fifty years, describes the hero's ultimate test as 'giving yourself to some higher end . . . When we quit thinking primarily about ourselves and our own self-preservation, we undergo a truly heroic transformation of consciousness.' The story theorist Christopher Booker, meanwhile, writes that 'the "dark power" in stories represents the power of the ego, most starkly personified in the archetype of the "monster" . . . This incomplete creature is immensely powerful and concerned solely with pursuing its own interests at the expense of everyone else in the world.' What Campbell and Booker are describing, of course, are the qualities of selflessness and selfishness – the human moral axis that came into being before we were fully human. Story's roots, it seems, run unimaginably deep.

Indeed, Booker also identifies a frequently occurring narrative archetype in which low-ranking characters 'below the line' conspire to topple the corrupt and dominating powers 'above the line'. 'The point is that the disorder in the upper world cannot be amended without some crucial activity taking place at a lower level,' he writes. 'It is from the lower level that life is regenerated and brought back to the upper world again.' Reading this, I couldn't help but recall the low-ranking chimps that conspire to fight their way to a place 'above the line'. It's not only the recognizably chimp–human pattern of fluid hierarchy that's still detectable in today's stories, it's also the design of the literary hero, story's model of ideal self. 'Stories present us with an ideal picture of human nature,' writes Booker. 'What we see endlessly recurring is the same equation: to reach the fully happy ending, hero and heroine must represent the perfect coming together of four values: strength, order, feeling and understanding.' The ideal self that is presented in our stories as the 'hero' is eerily similar the alpha chimp who, after struggling to the top, makes such a show of being powerful but also merciful and caring for the little guy.

If, as it appears, the deep foundations of story are tribal, then what has our Ancient Greek heritage added to the narratives we tell and the narratives we live? Of course, it's not possible to find a precise dividing line between them. But I couldn't help but sense the ghost of Aristotle in what the influential psychologist Professor Timothy D. Wilson has written about the narrative identities that psychologically healthy people adopt. These happy narratives feature 'a strong protagonist, a leading man or woman who takes charge and works towards a desired goal'. This goal should be freely chosen and we should be in control of our pursuit of it. 'The important thing,' he adds, 'is to pursue goals that give us a sense of autonomy, effectiveness and mastery.' All of which, to me, sounds suspiciously Greek.

But is it? One problem with thinking about culture's influence is that it's invisible to its wearer. When we hear ours described back to us, we can easily think, 'But that's not "culture" or a thing called "individualism", that's just the most obvious way of being human.' It's *natural* to see the world as made up of separate objects. It's *natural* to seek to challenge the authority of those in charge of us. It's *natural* to be competitive. It's *natural* to crave freedom. It's *natural* to fetishize celebrities and it's *natural* to strive to become one of them. Just as Aristotle believed that everything moves towards perfection, so any civilized group of people would surely, helplessly, flock towards these ideals. But this is not so. And we can know this for sure because, just as a Western self was taking form in Ancient Greece, in a place far over the eastern horizon, a very different kind of human was being made.

*

In the same era in which the Western self was being shaped by Aristotle and the individuals, a grumpy, sarcastic and pernickety thinker was roaming the other side of the planet, gathering disciples and trying to save the world. His was a land of war. For hundreds of years the rulers of the Zhou Dynasty had overseen their vast empire in relative peace. In contrast to the crags and islands of Greece, most of his country's population lived on great plains and amongst gentle mountains. These deep and wide horizons were a blessing for agriculture but were easy to conquer and rule centrally. They were also isolated: the inland people that farmed them rarely encountered foreigners or foreign beliefs. During the peaceful era of the Zhou, massive river-irrigation and water-conservation projects had been created. They were grand schemes that supported large farming projects and they all required the efforts of the many to succeed. Ancient China was no place for the ambitious individualist. Group harmony, rather than individual agency, was survival's favoured mode.

But by 500 BC, the glorious powers of Zhou had disintegrated. The region had descended into a chaos of massacre and conquest. It was into this mess that an astonishing and eccentric man, Master Kong, or Confucius, arrived. He became obsessed by the idea of bringing China back to its more harmonious days. He was an odd man – kind to some but foul to others, and rigid and pedantic. He made a point, in his dress, of avoiding silk lapels and cuffs of maroon. He didn't like to eat too much, and unless his dishes were prepared correctly and sauced appropriately, would refuse to eat at all. Manners were of great importance to Confucius. Coming across a young man sitting with his legs spread wide open one day, he caned his shins. But he himself could be magnificently rude, once feigning illness to avoid seeing a visitor named Ru Bei, then striking up a noisy song on a lute as poor Ru stalked off, just so he'd know he'd been snubbed.

We can have a good idea what the Confucian self looked like because it was written down, albeit some time after his death, by his followers in *The Analects*. This dutiful son of the plains could hardly have been more different from his proud, free and competitive contemporary in Ancient Greece. 'The superior man has nothing to compete for,' he's recorded as saying. 'But if he must compete, he does it in an archery match, wherein he ascends to his position bowing in deference.' He 'does not boast of himself', preferring instead 'the concealment of his virtue'; he 'cultivates a friendly harmony' and 'lets the states of equilibrium and harmony exist in perfection'. What Confucius calls the 'inferior man', meanwhile, could be a description of his showy Western counterpart. This is a person who understands not 'righteousness' but 'profit'. He is 'aware of advantage' and 'seeks notoriety' which causes him 'daily to go more and more to ruin'. Confucius believed that harmony could only exist among the people if they knew their place and stayed in it. 'The superior man does what is proper to the station in which he is. He does not desire to go beyond this.' And certainly not for personal gain. Writes historian Michael Schuman, 'Confucius expected people to do the right thing because it was the right thing to do, not because they'd get paid off at some point in the future.'

In his lifetime, Confucius failed. He wouldn't become truly influential for another two hundred and fifty years, when the warring period ended. The new rulers of the Han dynasty found his philosophy of deference and duty, which had been kept alive by generations of adherents, agreeable to their project of uniting and ruling the country. After all, Confucius had always preached that China should be led by a single emperor, a 'son of heaven', who had power over all. (Although, that being said, even the Han apparently needed a bit of convincing. Ancient historian Sima Quin noted of the Han dynasty's founder and future leader Liu Bang, 'Whenever a visitor wearing a Confucian hat comes to see him, he immediately snatches the hat from the visitor's head and pisses in it.')

The Han's ultimate embrace of Confucianism would end up changing the world for ever. Scholars such as Richard Nisbett argue that the flat and fertile landscape of China gave his ideas a kind of pre-destiny. In contrast to the Greeks, with their islands and city states, and their concomitant view of reality as a collection of individual objects, China's rolling, isolated, conquerable plains and hills produced a species of self that worked best as part of a group. It also resulted in them viewing reality not as a mass of objects, but as a realm of interconnected forces. For the Confucian everything in the universe was not separate, but one. It followed from this that they should seek, not individual success, but harmony. This perspective has a number of profound implications for the way the East Asian self experiences reality.

Ninety-five per cent of modern China belongs to the Han ethnic group and Confucius's influence still roars in places such as Japan, Vietnam and the Koreas. Incredibly, these differences, which are rooted in the physical landscape thousands of years ago, are still readily detectable in hundreds of millions of people alive today. That landscape has created distinct forms of self that live on, and not only see the physical world differently to us in the West, but have a different conception of what it is to be a human.

For the descendants of Confucius, reality is not a collection of individual objects but a field of interconnected forces. This means that East Asians tend to be more aware of what's happening in their environment: they'll see the whole picture, not merely its subject. They'll also understand that behaviour can be caused by the forces of the situation one finds oneself in, whereas an Aristotelian thinker, focused on individual objects and the powers they possess, is more likely to assume that a person acts as they do because they willed it. And so it's been found, in study after study.

Tests involving videos of fish show that Chinese people tend to put their behaviour down to factors in their environment, whilst Americans blame the character and wills of the fish themselves. Further studies, involving more videos of fish, found that students from Kyoto University were more likely to begin their report of the films with the context ('it looked like a pond') compared with those from the University of Michigan, who tended to start with the brightly coloured, fast-moving show-offy fish at the front. Although references to this 'focal fish' numbered about the same in all write-ups, the East Asians made over 60 per cent more references to objects in the background. Examinations of youngsters' drawings suggest these cultural differences develop gradually. Canadian and Japanese first graders draw pictures in similar ways, only for them to begin diverging a year later, when Japanese children begin including more pieces of information in their artworks and placing their horizons higher up, a tendency that's consistent with having a visual experience of the world that's more context-orientated, and is a feature of traditional Asian art that goes back centuries.

'It isn't just that Easterners versus Westerners think about the world differently,' Nisbett told me. 'They're literally seeing a different world. We've found that if you show people pictures for three seconds, the Westerners will look all over its main object and only occasionally make eye movements that drift out to the context. For the Chinese, they're looking constantly back and forth between the objects and the context. We track their eye movements every millisecond. This means they're able to tell you more about relationships in studies like the fish one. And that's why they're stumped if you show them an object by itself, out of its original context, and ask them if they've seen it before. Because what they saw was the object in context. The complexity of environments that Easterners can tolerate is much greater than it is for Westerners. I mean, the street scene in East Asia is just chaotic to us. And people say, "Oh, well, what about Times Square?" To which my answer is, "Yeah, what *about* Times Square?"'

The Confucian versus Aristotelian difference has also been detected in a study of newspaper reports. Researchers deconstructed stories in the *New York Times* and the Chinese-language *World Journal* about two mass murderers. They found the American journalists tended to blame flaws in the killers' characters – they suffered from a 'very bad temper' or were 'mentally unstable'. The Chinese reporters, meanwhile, emphasized problems in their external lives – one had lost his job, another found himself 'isolated' from the Chinese community. These findings were supported by interviews that found the Chinese more likely to blame life pressures for the killer's actions, with many believing that had his situation been less stressful, he might not have killed at all. The Americans' black-or-white, good-or-bad perspective, meanwhile, led to a greater conviction that the crime was inevitable.

As we've discovered, back in our tribal hunter-gatherer days, the thing that *all* human selves fundamentally want is to get along and get ahead. Everyone has this in common. When we're born, our brain looks to the environment to tell it *who we ought to become* in order to best fulfil this deep and primal need. What it's looking for is the model of the ideal of self that exists in its cultural surroundings. If the kind of self that will get along and get ahead most efficiently, in its cultural environment, is a freedom-loving, individualistic huckster, then that's who it'll want to become. But if it's a harmony-fetishizing team-player, then *that's* more likely to be the model it will aim towards.

This basic pattern – the local best-practice rules for success forming particular kinds of selves – has been found in other cultures too. Nisbett's team has studied three communities in Turkey's Black Sea region and found that those whose trades are based on networks of cooperation, such as fishermen and farmers, think more holistically than the shepherds, who rely chiefly on their own wiles. In the US, a study of male college students found that those from the southern states reacted more aggressively than their northern counterparts after they were shoved and called an 'asshole'. The psychologists predicted this outcome on the basis of there being an 'honour culture' in the south that grew out of the way their forefathers made their living: 'Herdsmen must be willing to use force to protect themselves and their property when law enforcement is inadequate and when one's wealth can be rustled away,' wrote the authors. 'In the Old South, allowing oneself to be pushed around or affronted without retaliation amounted to admitting that one was an easy mark and could be taken advantage of.' Evidence in support of their hypothesis was found in the southerners' saliva, which showed greater spikes in testosterone and cortisol (a hormone implicated in anxiety, arousal and stress), and also in their behaviour in a series of tests that immediately followed the insult. Not only were they more upset than the northerners, who were relatively unaffected, they were more likely to feel their masculinity was in question and to behave in dominant and belligerent ways.

A study led by Thomas Talhelm of the University of Virginia honed in on differences within modern China, finding that people from the south, where teamwork-intensive rice growing was common, were more collective in their thinking than those from the north, which favoured wheat. These differences, according to Talhelm, have long been acknowledged in China itself, where the stereotype is that those from the north are fiercer and more independent. In the US, of course, it's the southern states, with their independent herdsmen, that retain their reputation as being more violent.

Perhaps the greatest difference between the Aristotelian and the Confucian is in their tendency to be acutely conscious of being a part of a greater whole. The Asian self melts, at the edges, into the selves that surround it, whereas the Western self tends to feel more independent and in control of its own behaviour and destiny. Studies suggest, not only that Asians don't feel as in control of their lives as Westerners, but that they don't feel the need to be. Change is the function of the group, rather than the individual, their priority harmony rather than freedom. These deep substrata of thought can lead to startling differences above the surface. Amongst Chinese students, it's the humble and hardworking kids that are popular, whilst in industry shyness is considered a leadership quality. But other differences aren't so charming. 'The Chinese are willing to accept the idea of unjustly punishing someone if that makes the group better off,' said Nisbett. 'That's an outrage to Westerners who are so individual-rights orientated. But, to them, the group is everything.'

This manifested, too, in Confucian law, under which punishment for a serious crime extended to three generations of the criminal's family. I asked Professor Uichol Kim, a social psychologist at South Korea's Inha University, about a rumour I'd heard (and quickly dismissed) on a visit to Japan, that if you applied for a job, your family would be investigated. If your brother, say, had been to prison, you were likely to be rejected. This seemed so unfair as to be absurd. Was it true? 'Of course!' he said. 'If there's mental illness or a disability then you will not get the job. If it's in your marriage and/or your extended family also. So you hide it.'

The Asian self also reveals its porousness in language. In Chinese, there isn't a word for individualism (the nearest they have translates to 'selfishness'.) The term for 'human being' in Japanese and Korean translates as 'human between'. Most studies show that East Asians have lower self-esteem than Westerners. Richard Nisbett spoke of a Japanese friend who noticed that Americans always seemed to be looking for opportunities to build one another's self-esteem. 'If someone gives a talk, they'll say, "Great talk, man," no matter how bad it was. In Japan, they'd say, "Oh, I felt so sorry for you. You seemed so nervous." They don't regard it as an obligation to boost the self-esteem of other people. Of course, America set the scale for that nonsense. And, as in all these things, California set the scale for America. Basically, the further west you go, the more individualistic, the more delusional about choice, the more the emphasis on self-esteem, the more the emphasis on self-just-about-everything, until it all falls into the Pacific. I don't know if you're aware that California had a budget for increasing self-esteem?'

'I think I've read something about that,' I said.

'During that period, the Ann Arbor, Michigan, School Board, near where I live, had a debate as to whether the primary mission of their schools was imparting knowledge or raising self-esteem. Self-esteem won.'

If the Western self is a constantly unfolding story in which the individual hero strives towards perfection, it should hardly be surprising to discover the same pattern reflected in the stories we tell. Greek myths often involved brave heroes going on wild adventures in pursuit of dangerous monsters or bedazzling prizes. They mythologized the belief that great power could rest in the hands of one person who, if they struggled with sufficient courage, could change their lives, and the world, for the better. It seemed to me that, 2,500 years on, we're still telling those tales and living them. We guide our lives, and judge our worth, in accordance to their patterns. They form our ideas, not only of the proper shape of a life, but of who we ought to be. I wondered, if the Confucians didn't seek to struggle their way to personal prizes and glory, as we did, did that reflect in their mythic traditions? Did they have different kinds of stories to us? 'That's a very interesting question,' he said. 'You would think I'd have thought about that, but I haven't.'

I decided to ask Professor Uichol Kim. I was hoping to discover not just that East Asian stories are different, but that these differences reflected how their *selves* are different. If so, this would surely constitute yet more powerful evidence that self and culture really are symbiotic.

'In the East,' he told me, 'stories are different.' It isn't so much riches, nor the love of the maiden, nor the bravos of the many that tend to form the structure of their tales. It's harmony. This is the form many traditional Asian stories take: an incident such as a murder is recounted from the perspectives of several witnesses and then an event or twist takes place which, in some way, makes sense of them all. But don't expect this sense to be obvious. 'You're never given the answer,' he said. 'There's no closure. There's no happily ever after. You're left with a question that you have to decide for yourself. That's the story's pleasure.'

'And there's no implication by the storyteller as to which character's perspective was right?' I asked.

'They're all right. And they're all wrong. Of course!'

Likewise, in an Eastern form of story known as Kishōtenketsu, something happens, then something apparently unconnected takes place which makes us view the first thing in a new way. We're encouraged to search for the harmony between the incidents. 'One of the confusing things about stories in the East is there's no ending,' said Professor Kim. 'In life there are not simple, clear answers. You have to find these answers.' The Asian author often doesn't impart a simple lesson of wisdom in the telling of a tale. How could they when it's not possible for one hero, one author, to ever know the truth? 'How does anyone know the absolute truth? They can only tell what they know. You in the West see human beings as objects. But that is actually wrong. Human beings are subjects. A person is very egotistical. What I feel, what I see, is from my perspective. But someone viewing me can have a different perspective and a third person can have a third perspective. The truth is when all three perspectives are respected and combined. Then I arrive at harmony. But in the West it's right and wrong. It's simple.' This process of learning how to harmonize differing perspectives is what Asian thinkers mean when they talk of 'cultivation of the self'. 'It is the path of wisdom.'

Perhaps the most extraordinary and revealing difference in our storytelling comes in that inherently me-focused genre, the autobiography. What could be more obvious than recounting the tale of a hero from real life? And how else could you tell it than by reliving that hero's life, as if from their eyes, with them at the centre of the action, describing their decisions and views on the action around them? And yet, according to Professor Qi Wang, for nearly two millennia, there was 'hardly any real autobiography' in Chinese literature. And most of that which came to exist would be hardly recognizable to us. In China, accounts of an exalted person's life tend not to include their opinions or subjective facts about them. They are, instead, characterized by 'a total suppression of a personal voice'. Rather than being in the spotlight of the story, its subject is traditionally presented as a bystander, 'in the shadows'.

None of this is to suggest, of course, that there *aren't* any Eastern stories that centre on a 'hero' as we'd recognize them in the West. But according to Professor Kim, the hero's status is often earned a different way. 'In the West you fight against evil and the truth prevails and love conquers all,' he said. 'In Asia it's a person who sacrifices who becomes the hero, and takes care of the family and the community and the country.'

What unites the stories of our two cultures is that they're accounts of change. In the West we seek to bravely conquer the forces of change whilst in the East they seek a way of bringing them into harmony. But all stories serve the basic function of giving us insights into who we need to be in order to cope with the terrifying, ever-shifting world. In the memorable words of Professor Roy Baumeister, 'Life is change that yearns for stability.' No matter where we're from, stories teach us how to gain that stability. They are lessons in control.

And why does any of this matter? Because it brings us back to suicide. Since finding out about all the ways in which we model our selves and lives after the tales of our culture, I've begun to wonder if the perfectionist suicides might be stories that have broken. Back at the Suicide Lab, Rory had spoken of the suicidal mindset as having a sense of humiliation and defeat from which it cannot escape. 'You're trapped by life circumstances, you can't see a way out, or your job prospects aren't going to change and so on.' It reminded me of that Greek notion of a living hell – Sisyphus pushing a rock up a hill only to have it roll back down again for eternity.

If we're from the West, we'll be expecting our lives to follow a typical Greek trajectory: we'll be fighting our daily battles, winning our rewards, making our lives and maybe the world better, and moving steadily towards a state of personal perfection. Perhaps, I thought, the self begins to fail when we lose control of our narrative. Debbie and Graeme and Ross and Meredith and me – we'd struggled, but we hadn't come close to the state of perfection we felt culture was demanding of us. We'd become stuck, our narratives halted. Were we failed Greek hero stories? Was *that* our problem?

Whilst talking with Professor Kim, I realized I'd stumbled across a way of testing my theory. If suicides are failed hero stories, then did East Asian suicides follow Confucian plots? Did they kill themselves for different reasons, and did those reasons accord with their conception of what a hero is – a person who sacrifices and is loyal to the group?

The answer, I was quietly stunned to discover, turned out to be yes. In Confucian cultures, people's reasons for dying are often different from ours in just the way you might expect. In the East, it's those who neglect their duty to bring harmony to the group who are more likely to be considered failures. For women, in this highly patriarchal society, that can mean duty to family. 'If you cannot take care of your children, you kill your children and you kill yourself,' said Professor Kim.

'Does that happen a lot?'

'Yes. Recently a husband and wife decided to kill their children first and then commit suicide because there was no way they could take care of them.'

In China, it's not uncommon for corrupt male officials to kill themselves in order to halt a criminal investigation and let the family keep the dishonestly acquired bounty. In 2009 a former president of South Korea, Roh Moo-hyun, jumped from a cliff after being accused of taking bribes. 'He committed suicide to save his wife and son,' said Professor Kim. 'The only way he could stop the investigation was to kill himself.' In Japan, meanwhile, suicide by corporate and political leaders has long been thought honourable. 'A CEO of a company would consider his company as his family,' the anthropologist Professor Chikako Ozawa-de Silva told me. 'Instead of saying,

"Hi, I'm David," in Japan you say, "Hello, I'm Sony's David,"' she says.

'Even in social situations?' I asked.

'Even at very informal parties.'

In times of failure, this Japanese impulse to take professional roles so personally can be uniquely deadly. 'Suicide has been morally valorized for years, maybe centuries,' she said. 'It probably goes back to the Samurai. The idea is that by one individual taking his or her life, the honour is restored or the family member would be spared the shame. But that metaphor is so much extended, a CEO taking life makes sense to the Japanese. A CEO could say, "I'll take responsibility for the company," and take his life and the media and people in general would take that as a very respectable deed.' In a place where acceptance by the group is so important, its rejection can be disastrous to the self. 'This kind of fear might be universal, but in Japan it's emphasized,' she said. 'It's in people's minds all the time.'

So people in East Asia clearly have their own particular and serious problems with perfectionistic styles of thinking. Indeed, the suicide statistics can be devastatingly high over there. South Korea has, by some counts, the second highest rate in the world. Around forty South Koreans take their own lives every day, a rate five times higher than it was for the previous generation. One poll found that just over half of all teenagers had had suicidal thoughts within the previous year. Professor Kim believes much of this can be explained by the great miseries that have been unleashed by the country's rapid move to city life, and the resulting collision between collectivist Eastern and individualistic Western culture. 'Within one generation, 70 per cent of people who used to live in agricultural communities now live in cities. Confucianism was based on an agricultural culture where you know each other and you're supposed to care for each other. But in an urban city, it's very competitive and achievement-focused.' Which means the very definition of what it means to be a successful self has transformed. 'You're defined by your status, power and wealth which was not part of traditional culture,' he said. 'A Confucian scholar living on a farm in a rural village might be very wise but he's poor. We wanted to get rich.' The result has been a kind of amputation of meaning for the people. 'It's a culture without roots.'

He doesn't believe that recent glad news from China will hold. There, rates fell by an astonishing 58 per cent during the time of great movements from the countryside to the city. Professor Kim believes they're experiencing a 'lull' caused by this tide of hope. After all, South Korea saw similar drops when its economy was rapidly expanding. 'People believe when you're richer you'll be happier,' he said. 'When you focus on the goal you don't commit suicide. But what happens when you get there and it's not what you expect?'

*

By the time we were Greek, we'd come a long way. Our long evolutionary period as hunter-gatherers had given us tribal brains. The fluid nature of the groups in which we lived gave us a preoccupation with hierarchy and status. We wanted to get along *and* get ahead – urges that were frequently in conflict and inserted a hypocrisy into the core of the human self. We sought a reputation for selflessness, even if the reality of who we *really* were was more complex. We valorized those who gave for the tribe and punished those who selfishly took. We became prejudiced against members of other groups whilst unfairly favouring our own. Meanwhile, the interpreter in our heads narrativized our days. We tended to believe we were more moral, wiser and better-looking than other people. We developed personal narratives that gave our existence purpose and meaning. The atomized, adventuring economy of the islands and shores of Greece gave us a new individualist, rational, perfection-seeking sense of ourselves, and new stories on which to model our lives.

And then the cradle of the Western self fell. Ancient Greece's collapse led to years of division and war. The brilliant, inventive, freedom-soused world in which we found our essential cultural identity was no more. Old-fashioned tyranny robbed the Greeks of their most effective route to personal fame and success. Freedom was gone and, with it, hope and ambition. Their relationship with their gods had always been obstreperous, but now they turned on them in despair. The political philosopher Professor Sheldon Wolin writes, 'If the gods had been truly concerned with man's welfare, they would not have allowed the cities to disintegrate to a point where municipal life verged on a state of nature. If men could not trust in the divine agency of the gods, and if human perfection were no longer possible within the polis, the only conclusion seemed to be that man's fate was solely a personal matter.'

If our fate was now a personal matter, that meant turning the pursuit of perfection inwards. New thinkers, the Cynics and the Stoics, preached that human civilization was corrupt and that happiness lay in refusing its old lures. The perfect self was not one of fame and glory, after all, but one of pious virtue. The righteous man lived humbly and obediently. He trained himself to resist temptation. In order to protect our soul from the evil that was everywhere, we had to purge ourselves of the sinful excesses of our youth and become pure. And so we got down on our knees and we crossed ourselves and prayed.

Christianity eventually arose out of the fall of the ancient world. It thrived for centuries, forming and hardening around a social and economic landscape that was radically new. The Christian model of the ideal self was to last for so long because it suited the hostile realities of who we had to be, not least in the medieval era, if we wanted to get along and get ahead. It remains, today, the dominant religion of our people, and its moods and shapes live on even in those who have no faith in its stories.

This would be the next great chapter in the evolution of the Western self.

BOOK THREE

The Bad Self

It was an early evening in April when my taxi halted outside Pluscarden Abbey, its engine cutting out to silence. My search for the self had had me out of bed before dawn and now, hours later, I'd ended up in a remote Scottish valley, tired and lost. We'd first pulled up outside a low building just off the road that meandered through the landscape, but that had turned out to be the women's quarters. I was shooed away by a flustered lady in a woolly green jumper. My taxi drove on, at a respectful pace, past ploughed fields that ran with pheasants and rows of simple wooden grave markers, to the main abbey building, where it dropped me off before zooming back to twenty-first-century Elgin. I found myself in the quiet, utterly alone.

I'd come to Pluscarden with a quiver-full of missions. Firstly, I wanted to discover exactly how Christianity had changed the Western self. Next, I wanted to scan the monastery and its denizens for deep signals from Ancient Greece which, I hoped, might still be detectable if I paid sufficiently close attention. Finally, I was curious to see what fresh insights I could gather, from the men in robes I'd be living amongst, about the secret mechanisms of the human self.

I'd chosen Pluscarden to look for answers because the connection these monks had to this period in our history was extraordinarily direct. There'd been a monastery here since 1290. Then, as now, it was inhabited by Benedictines, holy men who lived by the rules laid down by a sixth-century Italian. Following the fall of the Roman Empire to the barbarians, a fourteen-year-old named Benedict became so disgusted with the paganism of Rome that he went off to live in a cave, thirty miles away, in Subiaco. Benedict was a superb hermit. So good, in fact, that word of his talents soon spread and he began to attract disciples (which might have been quite annoying, for a hermit). He went on to establish a monastery where, as an elderly man, he wrote down all the lessons he'd learned about running a band of monks. *The Rule of St Benedict* is a kind of how-to guide for anyone who wants to control a monastery or live in one. Today, the twenty-five monks of Pluscarden Abbey, 190 miles north-east of Glasgow, continue the rule's traditions by living a life that, they say, 'does not differ in any essentials' from the original.

Wandering around the abbey grounds, I saw empty benches, beehives along a distant hedge and a pile of torn guts in the grass, presumably left by a cat. Eventually I found a door at a side building that said it was for visitors on retreat. I knocked and stood back, looking up at it. Wondering about calling someone, I pulled out my phone. No reception. The wind barrelled down from pine forests in the hills above me and, huddling into my coat, I decided to approach the abbey itself. It was vast: pale stone with soaring arched windows, the structure itself the shape of a gigantic crucifix, shadows casting long across the grounds. Hearing footsteps behind me, I turned. And there it was, a vision from seven hundred years ago, a hurrying wraith, a phantom from the medieval, in a rough habit and sandals. He was a portly man, perhaps in his sixties, in off-white robes, large cuffs flapping around his arms, a rope around his middle, mud stains on his hem, pale-faced and slightly out of breath.

I followed him into the abbey as the failing light of dusk breathed low colours through the windows high above us. I entered a side chapel as he paced, apparently late, into the church that comprised the main body of the building. It was a cool space lined with pews that had a stygian dampness to it, its scent old paper, wax and incense, its climate guilt and dread. The monks, out of sight, sang a psalm in Latin, slow and sorrowful, its notes crawling into each other like an old man's fingers. It seemed perfect for its environment, this song of the stones. I picked up one of the booklets that had been left on the pew. It was a translation of the psalm I was hearing. What were these words, I wondered, that sounded so holy and gentle? 'The Lord at your right hand will shatter kings on the day of his wrath. He will pronounce judgement on the nations: corpses will be heaped up. He will crush heads across the wide earth.'

I put it down again.

The singing continued, on and on, wreathing itself around me. Eventually, curious again, I picked up another booklet: 'I confess to Almighty God and to you, brethren, that I have sinned exceedingly, in thought, word, deed and omission, by my fault, by my fault, by my very great fault.' It was as if I'd tumbled out of a time machine and been dropped straight into the overcast adolescence that followed the collapse of our narcissistic, ambitious, beautiful, noisy, avaricious, individualistic, sun-warmed Grecian youth. I felt incredibly depressed. 'Excellent,' I thought. I was in the right place.

With the psalms finally over, I was allowed into the men's accommodation block. I walked past a wooden statue of a glowering bearded man with a crooked staff, and into my room, which was called St Gregory. It was spare: a single bunk with a threadbare sheet, a waterproof underlay, a sink, a small desk with a Bible on top of other thumb-worn books of instruction and, above my bed, a heavy crucifix. I lay back, staring at Jesus's feet, and wondered about the strange sensations that had been blooming and roiling behind my experience for the last few hours.

Since I'd learned about confabulation, it had become almost second nature to watch my feelings as I was experiencing them with a kind of alienated squint. I was aware, and increasingly suspicious, of the separation between the things I felt and the voice that interpreted those feelings. We really are, as people sometimes glibly say, a mystery to ourselves. I'd wake up, every now and then, feeling unaccountably happy and whereas once

I'd have confabulated a reason why, I no longer bothered. It was the same when I'd wake feeling down. What's the matter with me? Why am I in this mood? I have no idea. So I'd just sit in it, gormlessly, like a dog in a puddle.

I'd come to think of my self as consisting of two separate parts: there was my interpreter, chattering away, often annoyingly, and then there were my emotions and drives, which were their own thing, churning in the background, effecting everything, sometimes overwhelming me. I'd noticed, too, the penetrating effect that something as situational as the weather could have not simply on my mood, but on how I interacted with people, how harshly I judged myself, how empathetic I felt towards people who happened to be in the news that day, on the person I actually was . . . my very self. For now, though, all I knew was that I was overcome by a feeling that was melancholy but at the same time safe and warm. Something about being around these Catholics was burrowing into me, as if I was regressing to a familiar and long-lost state.

My eyes flicked back to the crucifix above my bed . . . there he was, dying Jesus, looking just as he did in the churches of my youth, with his nearly naked perfect body – pecs, abs, pelvic v-line, his thighs and biceps straining with sexy agony. This 'son of God' might have been born in the Middle East, but in his nudity, appearance and spectacular display of *kalokagathia*, he was also as Greek as Hercules.

Just as I was about to fall asleep, I heard voices beneath my window. Curious, I climbed onto my desk and peered down to see two middle-aged men holding cups of tea, standing beside a bench. 'What I find weird is I'm not allowed to take the host,' a man with a white beard and a Geordie accent was saying to a younger, clean-cut man. 'I'm not Catholic,' he continued, 'but it's the same prayers and everything. If you came to my church, you'd be invited to take it.' He had a good point, I thought. The receiving of the host is the most meaningful part of the mass. It's when the magic happens, the climax of the story, the body of Christ magicked in a flatbread. It seemed a little hostile for the monks to be excluding him just because he wasn't Catholic. I expected the other man to back him up but all he did was stare at his shoes. 'Yeah,' he said, blandly. 'It gets complicated.'

Climbing back off the desk, I sighed. Tribalism. These monks were apes, as we all are, their white robes no less a gang-sign than the black suit John Pridmore was instructed to wear on the day of the showdown with Buller in the London pub. As I stepped down, I accidentally knocked the small pile of books from the desk. Bending to pick them up, I saw a slim red volume: *The Rule of St Benedict*. I took it to bed and began browsing.

I found it almost immediately, defined with halting clarity. Here it was, the new iteration of the Western self. And I couldn't believe how much it had changed. First of all, fun was out. 'Death is stationed near the gateway of pleasure.' So, it seemed, was individualism. 'Truly, we are forbidden to do our own will, for Scripture tells us: turn away from your desires.' Even the thirst for fame and glory was gone. 'A man not only admits with his tongue but is also convinced in his heart that he is inferior to all and of less value, humbling himself and saying with the Prophet: I am truly a worm, not a man, scorned by men and despised by the people.'

A despised worm? How had *this* happened? This dourly introspective self seemed so un-Greek as to border on the Confucian. Indeed, in a book by the former Pluscarden Abbot Aelred Carlyle, I went on to read that 'the main essential' of Christian life is to 'occupy the exact place God wills for us, and do our duty to that state, so that our lives become what He meant them to be.' I remembered what Richard Nisbett told me about individualism increasing the further you travelled west. Christianity, of course, had come from the East, albeit near rather than far. The sight of Jesus, up there, reassured me that the Christians hadn't scrubbed themselves *entirely* of their Greek cultural inheritance – this was an update to the self, not a replacement. But the change was clearly dramatic. And there was a troubling bass chord that rumbled beneath it all. Christianity seemed to have brought with it a feeling of self-hatred, a fetishization of low self-esteem. I'd recognized it, in the chapel earlier, and I could recall the feeling of it from my own Catholic upbringing.

I'd been raised in a Catholic household: Catholic school, fish on Fridays, church on Sundays, singing in the choir, serving as an altar boy, confession, confirmation, and so on. I couldn't remember ever having believed in God, but I also couldn't deny that Catholicism's crepuscular obsession with guilt and sin had become part of who I was. Sitting on the pew, earlier that evening, I'd been a jelly slipped back into the mould. The life lessons I'd internalized, as a naughty, guilty son and schoolboy, were such that I could never not be a Catholic. The faith had, in an actual, neurological way, become a part of me. The Catholic view of human nature formed the underlying texture of my daily experience – my implicit, pessimistic belief that human life is dangerous, unstable and corrupt. I found it hard to separate the hyper-critical inner voicethat narrates my low self-esteem from the voice of that church God who'd reminded us relentlessly, between the hours of 9:30 and 10:30 every Sunday morning, that we're all 'born sinners' and only total supplication to the powers of heaven could save us from the fire.

But now, for the first time, I had the beginnings of an understanding why the message of the medieval Church – with its introspective, scared, self-lacerating iteration of self – was what it was. Just as in the days of Ancient Greece and China, the kind of selves those long-ago Christians strove to be were ones best able to get along and get ahead. It was as if their brains had asked a fundamental question: who do I need to be in order to thrive in this environment?

In the brutal feudal reality of the Middle Ages, the successful self would have been compliant, hard-working and humble. Food was often scarce, the world was violent and there was no state to look after you. The rural many had to rely on powerful lords for protection and, often, for land to farm. Up to a tenth of the population were slaves, with the majority of the rest peasants or serfs who had to swear oaths of obedience to the lord of the manor and endure lives of debt, toil and servitude. Serfs were bound to their masters for life, often forbidden from owning property and had to secure the lord's permission to marry or leave the manor. I thought of St Benedict and his despised worms: 'Truly we are forbidden to do our own will . . .' The kind of human to thrive best in that environment would have been one who kept their head down and their ego in check, and would've been worshipful of the almighty powers that ruled them. And that was *exactly* the model of self I was meeting at Pluscarden.

At lunchtime, we were led to the refectory. The men on retreat were to eat at one end of the spartan room whilst the monks lined up by long tables that lined the white walls. The ritual of the meal was remarkable. They stood and sang grace, then the abbot knocked on the table, they raised their cowls over their heads and a monk, high up at a carved wooden lectern, read from the *Rule of St Benedict*. The abbot knocked once more, and they sat, simultaneously, in preparation for eating, which took place in silence. When the food was finished they cleaned their bowls and the ritual began again, at all times led by the abbot.

What I was watching was a ritualized display of mimicry and deference towards the abbot from the subordinate monks. This, in a sense, is how we all behave in the presence of people that our brains have identified as leaders. The difference is that, when we do it, it's automatic and mostly unconscious. We're not aware we're behaving in this way – and neither can we help it. To understand how this process works is to discover another critical way in which culture spreads out and infects us.

*

So far, the researchers who study the 'geography of thought' have revealed the extent to which *where* we are shapes *who* we are. The sort of person we become depends, to a great extent, on who we need to be in our particular environment. But that's not to say that individual men and women don't have the power to change us too. To argue this would be to deny the effect that Jesus, Aristotle, Confucius and all the other remarkable cultural leaders we're to meet on this journey have had on us. The fact is that when we're working out who we need to be in order to get along and get ahead, we're not just taking our information from stories. Being tribal animals, we're also constantly scanning our environment for *people* who seem to have, in some way, mastered the secrets of a successful life. The ideal self we're looking for doesn't only exist in fiction and gossip, it's also right there in front of us. And these people can be a powerful source of influence. The psychologist Professor Joseph Henrich writes that the 'cultural learning' that comes from those around us 'reaches directly into our brains and changes the neurological values we place on things and people, and in doing so, it also sets the standards by which we judge ourselves.'

Our brains identify these leaders by being alert to various 'cues' that they, and the people around them, display. A basic cue we look for is 'self-similarity', for the straightforward reason that we're more likely to learn salient things by deciding to follow people who are like us in some fundamental way. (Our instinct to be drawn to and mimic those who are similar to ourselves is, sadly, yet another way that we're automatically tribal.) Another cue is age, which is especially important for children. Physical dominance is a cue that can be traced back to our primal ancestors and was, of course, John Pridmore's favoured method of exerting influence. But we also hunt for two more mercurial qualities that do much to explain not only the way in which individuals end up having outsized effects on their culture, but the often barmy world of celebrity we inhabit today. These cues are success and prestige.

Research suggests that we start mimicking people who we see displaying competence when they're completing tasks at around the age of fourteen months. As we grow up these 'skill cues' begin to take on a more symbolic form, as 'success cues'. In our hunter-gatherer pasts, it would've made sense to copy the actions of the hunter who wore many necklaces of teeth made from his kills, for example, as his success cue demonstrated high competence. It seems likely that designer clothing, expensive manicures and fast cars are today's equivalents of these attention-magnetizing displays. Success cues impress because of how our brains have evolved. You might argue that an investment banker's Ferrari doesn't signal any kind of excellence you're interested in – or, indeed, any kind of excellence at all. That, sadly, is beside the point. This behaviour is automatic and unconscious. It just happens. (And if you *are* somehow immune to the cues that come with wealth, there'll surely be another set of success cues that have an equally powerful and largely hidden effect on you.)

But we don't just rely on our *own* sense of who's skilful and successful when we work out who to copy. As a highly social, groupish species, we tend to look at who *other people* consider worthy of attention. We'll note, for a start, that these men and women have plenty of admirers. Like St Benedict, the world's most popular hermit, everyone seems to be somehow drawn to them. 'Once people have identified a person as worthy of learning from,' writes Henrich, 'they necessarily need to be around them, watching, listening and eliciting information through interaction.' This chosen person, and the people around them, will then begin to show 'prestige cues'. The chosen person's body language and speech patterns will show differences. Others will defer to them in myriad ways, conversationally and with eye contact. They might give them gifts or help them with chores, they might curtsy or bow or, as the monks do to the Pluscarden Abbot, copy them in overt and ritualized ways. Without realizing it, they'll often mimic their body language, mannerisms and vocal patterns.

One of the more sinister ways we mimic and signal our deference to cultural leaders happens entirely outside of our conscious awareness. The human voice contains a low-frequency vocal band of 500 hertz that was long considered useless because, when the higher frequencies are filtered out, all that's left is a deep information-less hum. It's since been discovered, though, that this hum is actually 'an unconscious social instrument'. The dominant person in a social situation tends to set the level of the hum and everyone else adjusts theirs to match it. Analyses of twenty-five CNN interviews given by Larry King found that he changed his hum to match George Bush and Liz Taylor, signalling his deference to them. However, Dan Quayle and Spike Lee adjusted to match him. Perhaps tellingly, in the more prickly interviews, such as that with Al Gore, neither party accommodated the other.

We're naturally attracted to prestige cues, and begin to follow them early. A clever study by a team including Henrich had pre-schoolers watch a video of two people using the same toy in different ways. As they were playing with it, two bystanders entered the room, watched the first person, then the second person, then 'preferentially watched' just one of them. 'The visual attention of the bystanders provided a "prestige cue" that seemingly marked one of the two potential models,' writes Henrich. Afterwards, children were thirteen times more likely to copy the way the prestige-cued person had used the toy.

The way we respond to these cues, by automatically deferring and copying, seems to explain one of the stranger facets of modern celebrity culture. It's why, for example, the ex-boxer George Foreman's view on contraptions that grill meat can apparently be taken seriously, at least unconsciously, by many of the *hundred million customers* that have bought them. If someone gives out prestige cues, we're naturally triggered into this behaviour, especially if that person is part of our perceived in-group. The mind isn't worried about whether it actually makes sense – whether this person's sphere of excellence is actually useful in judging the product they happen to be selling – it's just a dumb mechanism picking up on cues and behaviour.

All of this leads to a phenomenon that's sometimes known as the 'Paris Hilton effect'. Because we're wired to direct our attention towards the people who are already the subject of attention, we'll sometimes be drawn to people in the media without really knowing why. But our being drawn to them makes the media focus on them even more. We then attend to them more, then the media attends to them more, and then there's a runaway effect, a feedback loop, in which the status of an essentially nondescript person becomes madly amplified.

So we copy people. We're helplessly drawn to them. We identify the ones who seem to know best how to get along and get ahead, we watch them, we listen to them, we open our selves to their influence. And then we'll often internalize the things they've taught us. They have become absorbed into our model of the perfect self. They are now part of us. And, so, culture spreads.

*

After lauds, at 4:30 the next morning, I became distracted by a quiet path that led through the wooden grave markers. Curious, I followed it. Behind an on old stone wall I came across a Londoner called Robert. Pale, in his mid-forties with thinning curly hair, small round glasses and a blue raincoat, he told me he was staying at Pluscarden because he was considering becoming a monk. 'It's scary,' he said. 'But then you think, "Is it just the Devil trying to put me off?" If you have faith, you shouldn't be scared of anything.'

'Why is it scary?'

His voice dropped. 'There's no getting away from it,' he said. 'You come here to die.'

'*Die*?'

'Well, it's like the old self gets killed off and they replace it with the Holy Spirit.'

'I was thinking about that,' I said. 'There's probably not a lot of opportunity to sin in a place like this. You're going to become a better person almost by default.'

The look on his face told me I wasn't getting it.

'Living for yourself is living in sin,' he said. 'What they're doing here is *opus dei*. The work of God.'

'But aren't you worried that doing it every day will get a bit boring?'

He looked exasperated. 'But that's the *point*.'

Back in the grateful safety of my room, I puzzled over what Robert had said. The point of being a monk is to be bored? I found a reference to the '*opus dei*' he'd spoken of in a book I'd picked up in the gift shop. He was right: it meant 'the work of God' and it was described as 'an unceasing round of prayer, chant and ritual'. Unceasing! On an information board back in the abbey, I'd read that the monks spend five hours in church, three and a half hours doing spiritual reading and four hours at manual labour, 'leaving about 30 minutes free time'. They weren't engaged in outreach work with the community, as I'd assumed. Their function was not to help people. They weren't allowed possessions. They weren't even allowed to keep the names they'd been born with. They just seemed to be eating, sleeping, waking up, dressing the same, thinking the same, praying from dawn until dusk, parts in a clock that never unwound.

This, of course, goes against the logic of everything I'd been learning about people. 'A protagonist is a wilful character,' writes the story scholar Robert McKee, and the same goes for people living psychologically healthy lives. It's impossible to really understand the human self without grasping how important the things we do with our lives are to our emotional and perhaps, even, physical well-being. We've learned already that the mind makes us the hero of our lives. But heroes *do* things. To make a successful story, a self needs a mission. It needs a plot.

Humans just can't stop themselves being the causer of effects. Psychologists describe the 'effectance motive' – the drive to manipulate and control elements in the world – as 'almost as basic a need as food and water'. When people are left floating in darkened salt-water tanks with their eyes covered they experience what's known as 'stimulus-action hunger', which they'll seek to soothe by rubbing their fingers together, say, or making waves in the water. One clever study saw 409 people stripped of their phones and left alone in a room, for up to fifteen minutes, with nothing to do – except use a machine to give themselves electric shocks that were so painful, participants said they'd pay money not to experience them again. 67 per cent of men and 25 per cent of woman were sufficiently discomforted by this they began shocking themselves. The researchers concluded, 'Most people seem to prefer to be doing something rather than nothing, even if that something is negative.'

It seems brain and body respond in positive ways when we are actively making progress with our lives, pursuing the plots that give them meaning. The neurobiologist Robert Sapolsky has argued that the brain's dopamine reward system, which guides our behaviour by giving us little druggy hits of pleasure, is more active not when we seize the prize that we're after, but when we're in pursuit of it. Meanwhile, work by geneticist Professor Steve Cole and his colleagues is beginning to suggest that our physical health might improve – risk of heart disease and neurodegenerative disorder going down; antiviral response going up – when we're successfully engaged in pursuits that are meaningful, a state that Aristotle called eudaemonic happiness. 'It's kind of striving after a noble goal,' he told me.

'So it's heroic behaviour in a literary sense?' I said.

'Right. Exactly,' he said.

Further studies have found that people with a greater sense of purpose, and more likely to agree with statements such as, 'Some people wander aimlessly through life, but I am not one of them,' actually lived longer than others, even when factors such as age and well-being were controlled for. These goals that we purposefully pursue have been the subject of many decades of research by the psychologist Professor Brian Little. He calls them 'personal projects' and has examined tens of thousands of them in thousands of participants. He's discovered that people are typically engaged in around fifteen at any one time. Whether they be mundane, such as teaching the dog to sit, or meaningful 'core projects' such as trying to rid the world of racism, he believes they're so essential to our sense of self that they actually *are* our self. 'In many respects we are our personal projects,' he told me. 'We are the things we're doing.'

Little has found that in order to bring us happiness, a project should not only be meaningful, we also must have some sense of control over it. Traditionally, heroes in fictional stories are ultimately successful in their struggles to get what they want, and that's how *we* should feel too – that, no matter how tough things get, we're making some kind of progress. When I asked Little if the idea of the core project speaks to an existence close to that of literary heroes battling to make their lives better through the archetypal three-act narrative structure of crisis–struggle–resolution, he replied, 'Yes. A thousand times yes.'

In order to be happy, then, we really ought to be living our lives as story. We should have a goal and feel like we're at least somewhat successful in our pursuit of it. Suicide is what happens when the progress halts, robbing us of our hero status. But if this is so, then what was the monks' story? Where was the fight and the hope? What form did the struggle of their lives take, in such a way that they could muster sufficient motivation to carry on, hour after hour, year after year, until they ended up beneath one of those wooden crosses out there by the drive? It was as if the monastery had been designed, specifically, to eradicate the most fundamental desires of the self. These men were stuck. And if the ideas I'd been pursuing were correct that meant they should be at risk of being suicidal. I mean, they did seem rather grumpy. But still . . . Picking up Abbot Aelred, later that day, I noticed the following observation about 'materialists' who don't believe in God: 'Spiritual things are to these people a sheer waste of time, and to them the Consecrated Life must indeed seem a dismal affair, bound by arbitrary rules, a barren, wasted life.'

'Amen to that,' I thought.

*

Father Martin was nineteen when he first wandered up the long drive to Pluscarden. He was convinced the monks were just going to throw open the doors in delight at his arrival and begin measuring him up for his habit. But he wasn't even a Catholic. The man who received him, Father Maurus (who'd one day bring Pluscarden a moment of media fame by vanishing, never to be discovered), instructed him to go away and live a little. Eighteen months later Martin triumphantly returned. 'Here I am!' he said. 'I'm a Catholic now!'

He lasted a week.

'I couldn't take it,' he told me.

'What couldn't you take?'

'I couldn't even tell you.' He thought for a moment and then burst into a smile. 'The whole thing! It was everything that was going on here!' He threw his arms up and roared with laughter. 'I mean, *what's going on?*'

I was talking to Father Martin in a small room that was almost literally bare. There were three simple chairs on the carpet-less floor, and nothing on the white walls except a crucifix and a double power socket. He was sitting in a plastic seat with his legs stretched out in front of him, his hands resting on his belly. He was sixty-six years old, with a mild Fife accent and had curious hair: it was as if he'd shaved off his sideburns but finished an inch too high, so now precise rectangular wedges of scalp were exposed above each ear.

He told me that a few years before his first attempt at joining the brotherhood, his parents' business of forty years had gone bankrupt. 'My parents didn't discuss it with me,' Martin told me. 'But tension and anxiety communicates itself.' His father was a 'tough guy' who'd had a distinguished record in World War Two. 'He'd come up from the ranks to become an officer, which wasn't all that common. He was very, very, very proud. He found the shame of the failure almost unbearable.' So unbearable that he decided that the family had to leave Fife for ever. 'He thought he had to get as far away as possible,' said Martin. 'It could easily have been Australia. It turned out to be Dufton.'

The bankruptcy and its fallout caused a rift in the family that Martin, then a teenager, found greatly distressing. Around this time, he became bothered by nihilistic thoughts. 'I got to the stage of thinking, "Why do anything? What's it all about?" I was looking around religion, reading books about yoga and Buddhism. I went to this thing called Subud, which is a religion that started in Indonesia.' He turned to philosophy for answers to the great questions, but each argument just seemed to dissolve into discussions of semantics. Fretting about the problem of free will, he saw the great thinkers concluding the whole thing was just a mistake about how we use words. 'I thought, "This is rubbish!"' Just as he was tilting towards despair, he mentioned, vaguely, to his father that he'd been thinking about the monasteries. 'He said, "Well, son, the only monasteries I've ever heard of are Catholic ones. Why don't you go and talk to the priest at the church over the road?" So I did. The priest was very matter of fact. He said, "Well, there's a monastery nearby. The beds are hard and the food's not good." And that was this place.'

The day after Martin left Pluscarden, following his second failed attempt, he found work on the night shift at a fish-freezing factory. 'You froze these plates of smelly fish and the equipment ran on ammonia gas and there'd be leaks so the smell of ammonia and half-rotten fish would be everywhere and you were bitterly cold. I thought, "Gosh, in twenty-four hours I've gone from heaven to hell."' His final, successful attempt wouldn't come until 1994, after a long stint as a parish priest. He's been at Pluscarden, now, for twenty-one years.

I was curious to find out what it was that had drawn him, so insistently, to the cloistered life. 'Was it a feeling? A voice?'

'There was no voice,' he said. 'In my experience, when you talk about the call of God, it's just an impulse to do something. You don't think, "Can I do it?" or "I should do it." You just go and do it.'

'So it's before the thoughts even begin?'

'Yes.'

'Like a magnetic pull?'

'Yes, exactly.'

Just as lack of faith was a facet of my unconscious, and not the result of any process of conscious reasoning, so Martin's faith was born beneath the speaking part of his mind. It was a force within him, a truth that wasn't thought, but felt. All the rest was confabulation. But faith wasn't enough to explain his presence here. It seemed to me that there must be something particular about Martin's psyche that was feeding off this life of deadening repetition, this regime of circularity and stuckness that should, in theory, be killing it. I asked him if there were many instances of men failing to endure it. 'In my time, there have been lots who've tried the life and an awful lot who haven't remained,' he said. 'I can think of two that were asked to go,' he said. 'Obviously there are procedures if there's friction, but one of them seemed to have friction with everyone. Once or twice it even led to blows. The other one wouldn't accept large chunks of the teachings of the Catholic Church.'

'What are these "procedures" for friction?'

'Well, to take an example, if you see a senior member of the community is angry with you, then you've to go flat, prostrate in front of him. You have to take the initiative and do something placatory, even though he's the one that's annoyed.'

'You're subsuming your ego?'

'That's right, yes,' he said. 'Then of course there's a whole disciplinary thing. The Rule sets out what you do for certain misdemeanours during worship and that's carried on today in most monasteries. It's a structure of asking you to humiliate yourself, as it were, to acknowledge that you were a wrong-doer. If you make a mistake, you have to kneel down. You're apologizing to the Lord but you're also acknowledging your neighbour, who you've irritated. If it's a more serious thing, you've to go out in front of the abbot and kneel down.'

'And you do this after the mass?'

'During,' he said. 'As soon as you've done it. But that doesn't necessarily quell the anger.'

'And then what do you do?' I asked.

'Well,' he shrugged. 'I mean, monastic life carries on.'

Martin spent fourteen years growing veg out in the garden. That's more time than he spent being a parish priest. *Fourteen years.* I don't know whether he was feigning or not, but when I asked him if the life ever got boring, he appeared taken aback. 'Er . . . um . . . bored?' he said, apparently baffled by the very suggestion. 'It hasn't been a problem so far. Even in the monastery you can't know what lies before you, due to infirmity and so on. It might be that you can no longer get to choir and you're just in your cell.'

It wasn't the most convincing example of life-enriching novelty I'd ever heard.

'So you're singing the same psalms again and again and you're growing veg for fourteen years and you're never bored?'

Martin looked into the corner of the room, apparently searching for an answer that might prove helpful to me. 'You . . . um . . . you can become *distracted,*' he offered. 'You can get so into something that you lose who it's for. Unfortunately that can go for even the worship. I sometimes get intrigued by the mechanics of singing. That often distracts me from the praise aspect. It might even, but hopefully doesn't, nullify it totally. One of the main dangers for those who really get their teeth into the mass and the asceticism and all that is that the Devil has a go at the vanity and vainglory aspect.'

'So you're thinking, "I'm the best monk"?'

'Exactly!' he said. '"I fast more than anybody else." And when it's my turn to cook and someone says, "That was really good," I think, "Oh!"'

'And that's the Devil?'

'Yes! Yes! He's nullifying the whole thing!'

'With competitiveness?'

'Yes! Yes! You see, you take yourself with you wherever you go. You can't leave yourself behind. The whole of the monastic life is like a hospital, trying to heal these defects of character.'

So these monks *had* dedicated their lives to the pursuit of the perfect self just as the Greeks had – it's just that they differed in their understanding of what the 'perfect self' was and how you should get it. For the Greeks, the inherently valuable individual strove to become perfect in order to win prizes, fame and boons for the community. The Christians had taken that struggle and turned it inwards. For them, it wasn't about Olympic glory or being the greatest potter, or the highest leaper of bulls, it was about a continual battle – with prayer, self-denial, flagellation – to make their *inner* selves better. What the Christians had added was interiority. Becoming a hero, now, meant being perfect inside *and* out.

This is where low self-esteem gets built into the core of the machine. For Aristotle, a person had innate potential and was naturally moving towards perfection. But for the Christians, a person was born in a state of sin and falling towards hell. God, not the individual, was where perfection lay. This meant that a person wanting to become more perfect would have to engage in a constant war with themselves – a war, not with forces out in the world, but with their own soul, their conscience, their mind and thoughts. And because perfection *only* existed outside the human realm, that struggle would always be hopeless. The Christians had given the Western self a soul, and then begun to torture it.

Of course, this preoccupation with the state of our inner selves – its moral cleanliness, its strength, its 'peace', its worthiness – is still an enormous part of our culture and daily experience, regardless of our faith, not least in its manifestation in the multimillion-dollar self-help and wellness industries. I have, in many ways, experienced my own years of conflict with low self-esteem as a dualistic, biblical war against dark forces that lay within me. Indeed, when we struggle to improve ourselves, in some way, we often frame it as a battle with our faulty inner selves. We talk of our failures as our 'demons', who are part of us, and who we must fight. *By my fault, by my fault, by my very great fault . . .*

None of which is to say that the Greeks had no interest in moral goodness, of course, or that there were no medieval Christians who wanted to better their economic circumstances (indeed, the monasteries were pioneers of an embryonic form of capitalism). But it was a shift in emphasis – and a major one. It changed who we thought we were, who we wanted to be, and the method by which we got there, for ever.

And there was, I discovered, at least one more critical way in which the Ancient Greek self lived on in the Christians of the Middle Ages. As unlikely as it may seem to some, Christian belief has always had the progress and reason that's characteristically Greek embedded within it. Unlike the holy books of Islam and Judaism, say, the New Testament was always understood not to be the literal word of God, but recollections about Jesus 'according to' his disciples. This left a gap of understanding to be filled by study and debate. As the social scientist Professor Rodney Stark has observed, where the Qur'an confidently asserted itself to be 'the Scripture whereof there is no doubt', St Paul admitted that 'our knowledge is imperfect and our prophesying is imperfect'.

That meant that, in order to really understand what God wanted, we had to analyse and reanalyse and then argue about the disciples' recollections. In the fifth century, St Augustine wrote that Christians should 'approach together unto the words of Thy book, and seek in them Thy meaning'. The idea was that the truth was buried in there, somewhere, and that the more we studied, the more we'd reveal. Writes Stark, 'From the early days, the Church fathers taught that reason was the supreme gift from God and the means to progressively increase their understanding of scripture and revelation. Consequently, Christianity was oriented to the future, while other major religions asserted the superiority of the past.' This is why Islam and Judaism are known as 'orthoprax' religions: they're concerned with following the letter of the written Holy law; with correct (*orth*) practice (*praxis*). Christianity, meanwhile, is 'orthodox', concerned with correct (*orth*) opinion (*dox*). It could hardly be more Greek.

Although, it should go without saying, there are notable exceptions on all sides (there are Christians, for instance, who believe the Bible to be the literal word of God. We call them creationists. They're not people who could comfortably be described as 'future-orientated'), the emphasis, according to Stark, was always 'on discovering God's nature, intentions, and demands, and on understanding how these define the relationship between human beings and God'. In place of learning and reciting holy law, Christian preachers are more likely to use excerpts from scripture to construct an argument about right and wrong. The priest, the vicar, the pastor, up there in the pulpit, is Aristotle in the Lyceum, using debate to reveal deeper truth. It was this faith in the struggle of reason to make a better future that led, in the twelfth century, to the foundation of the university system. As much as it might choke many atheists to concede it, it seems we owe those restless believers a lot.

It would certainly have choked me, back when I was a teenager, furiously rebelling against my parents' Catholicism. At Pluscarden, I'd been taken aback by the extent to which I'd been reminded of how it felt to be a boy, except instead of trying to please Dad and Mum, it was Almighty God and the Virgin Mary. If I'd felt like a child again, perhaps that's partly because childhood is what the monks had unconsciously recreated for themselves. 'I suppose there are two ways of looking at all the routine,' I said to Father Martin. 'You could consider it deadening. But you could also see it as reassuring.'

'Yes, yes, yes, yes, yes,' he replied. 'There is a reassuring aspect to it. Yes, yes, I wouldn't deny that. Benedictines make a vow of stability, of staying put. It took me three attempts and over twenty years to get here and, when I did, I made a vow of stability. And then they asked me to go to a monastery in Ghana!' He let out a great bellow of a laugh. 'That was the biggest crisis I've ever had.' A serious, distant look fell over him. 'I've been to Ghana six times now.'

If the job of the self is to give us a feeling of control over our unpredictable selves and our chaotic environment, it sounded as if there was something in Father Martin's character that needed this feeling especially badly. He seemed like a person who was unusually fearful of change. I wondered if that was why he'd been so powerfully drawn to this life of maximum predictability.

I wondered, too, about the ultimate point of his lifetime of self-obsession. During my week at the abbey, I'd detected the presence of the Greeks in Christianity's inherent belief in reason and progress, in its struggle towards personal perfection and in its near-naked, *kalokagathia*-filled images of its top celebrity, Jesus Christ. But I also had a growing suspicion that Christianity and the monastic life might be more baldly, Greekly individualistic than even that. In fact, they might even have surpassed them. For thinkers like Aristotle, the ultimate point of self-pursuit was that a person would be of more value to their community. But didn't these monks essentially believe that by doing the right thing they'd earn a fabulous future reward for themselves? Could it really be true that beneath all the outward humility and subservience there was a cold, steel heart of self-interest?

'Are your activities here a kind of attempt to prepare for the afterlife?' I asked, before I left.

'Yes, indeed.' He pointed upwards. 'It's a school for there.'

'So you struggle and sacrifice here in order to obtain a better life in the future?'

'That's right,' he said. 'Yes, yes, yes.'

*

If who we are is, to a significant extent, our culture, and if that culture is partly composed of the arguments, discoveries, feuds, prejudices and mistakes of dead men and women, then those arguments, discoveries, feuds, prejudices and mistakes live on, in some form, inside of us. We've internalized them. They've changed who we are. They're recorded in our brains as patterns of synaptic connections. They *are* us.

One of the strangest chain of events that changed who we are, in the West today, began in a cramped room in nineteenth-century Moravia. It was in that small space that a boy called Sigi lived, with his rapidly growing family. Along with his siblings, he was probably present when his parents had sex. He was present, too, when his brother died. Julius had been born just eleven months after Sigi and, in doing so, had robbed Sigi of the golden blessing of his mother's attention. But even Julius's death didn't return his beloved mother to him; before he was ten, she would have six more babies. Whilst Sigi developed a resentful, jealous hatred for his father, his maternal care came mostly from a nursemaid, Monica, who taught him about God and hell and would help him fall asleep, it is thought, by quietly stroking his cock.

It was all so confusing for the clever, earnest little boy. How could it not be? Consider the tangle of his immediate family. Back when Sigi's parents married, his father had been twenty years older than his mother. In fact, he'd already been a grandfather, with two adult sons from a previous relationship. This meant that one of Sigi's half-brothers was the same age as his mother, and the other a year older. These 'brothers', then, were old enough to be his father. Meanwhile, one of his dad's grandchildren was the same age as him, whilst the other was a year older. No wonder he grew up feeling muddled. The ages and generations and roles were all mixed up. Perhaps because of this, the concepts of sex and arousal weren't quite in their proper place. When he was seventeen, Sigi had a crush on a fifteen-year-old called Gisela. At the same time, he had a crush on Gisela's mother. He also nursed a crush on his own mother.

It was a story from the foothills of the Western self that would change Sigi's life: a tale from Ancient Greece. In 1873, as part of his final school examinations, Sigi read *Oedipus the King*. It was a kind of detective story with an ingenious plot: the ruler Oedipus had been told that a plague would only cease when the previous king's murderer had been revealed and expelled. Oedipus vowed to do the job. The man he was looking for, a prophet predicted, would turn out to have a muddled family background: 'He shall be shown to be to his own children at once brother and father, and of the woman from whom he was born, son and husband.' Over the course of a stunningly dramatic plot, Oedipus slowly realized that the man he was looking for was, in fact, himself. Years earlier, in a road-rage incident at a crossroads, he had unknowingly killed his father, King Laius. After that, he'd accidentally married his mother.

There was something, *something*, about this story that left Sigi profoundly moved. The emotional response it detonated in him was so enormous that he concluded it must possess a particular and special quality that resonated deeply with the very essence of human nature. But what could it be? He studied the text. He went to see it performed, to rapturous applause, in Paris and in Vienna. As he grew up, he became fascinated by the power of stories, and the mystery of how they moved such great masses of people. And no story obsessed him more than *Oedipus the King*.

From what we know now about the left-brain interpreter we can wonder if the play acted as a kind of ready made confabulation for Sigi, a makes-sense story that fitted so perfectly over the traumatic, shameful muddle that happened to exist in his head, and made him feel better about it. As historian Professor Peter Rudnytsky has observed, 'The coincidence between his biographical accidents of birth and the Oedipus drama is staggering.' And as the man himself was to write, 'I have found love of the mother and jealousy of the father in my own case too, and now believe it to be a general phenomenon of early childhood . . . If that is the case, the gripping power of *Oedipus Rex* . . . becomes intelligible.'

As a student, at the University of Vienna, Sigi would wander the Great Court, amongst the sun and shadows of the arcade, perusing the busts of former professors. He'd imagine himself among them one day. He could see it . . . There would be his name, in full . . . Sigmund Freud . . . and what would the inscription say? Ah, yes, it would be a reference to his hero, of course, the searcher of truth from Ancient Greece, the killer of the hated father, the lover of the beautiful mother, Oedipus the King, who, as the playwright Sophocles had written, 'knew the famous riddles and was a man most mighty'. It was a fantasy that was to come true, almost. Freud was to become the founding father of psychotherapy, his life's mission to unbury, detective-like, the hidden forces that writhed in the human unconscious; urges that were often violent and perverted.

Central to his ideas was Oedipus's tale. Childhood attraction to the mother and murderous hatred towards the father was, he decided, not just the experience of himself but the 'inevitable fate' of all of us. 'Every new arrival on this planet is faced by the task of mastering the Oedipus complex,' he confidently declared. 'Like Oedipus we live in ignorance of these wishes, repugnant to morality, which have been forced upon us by Nature, and after their revelation we may all of us well seek to close our eyes to the scenes of our childhood.'

*

In his assumption that he was just like everyone else, and everyone else was just like him, Freud was not alone. It's well known by modern psychologists that most people tend to significantly over estimate the extent to which others share their feelings and beliefs. Professor Nicholas Epley, who's studied this effect, writes that, 'Brown-bread lovers think they are larger in number than white-bread lovers. Conservatives tend to believe that the average person is more conservative than liberals do. Voters in both sides of an issue tend to believe that those that didn't vote in an election would have voted on their side. And when it comes to morality, even those who were clearly in the minority nevertheless tend to believe they are in the moral majority.'

Some of Epley's own experiments have focused on the extent to which people believe that *God himself* shares their perspective. When participants in a brain scanner were interrogated about God's views, and then their own, there was no observable difference in brain activity. Such tests are sometimes controversial, with sceptics doubtful that many solid conclusions can be drawn from them. But these particular findings have been supported by non-brain studies, in which people's own views have been seen to change in concert with their imagined views of God. 'When others' minds are unknown,' writes Epley, 'the mind you imagine is heavily based on your own.'

But with this very ordinary error, Freud managed to redefine our concept of original sin. With our 'wishes, repugnant to morality . . . forced upon us by nature', he theorized that much of humanity's inner misery came, not from the Devil's temptations, but from the monstrous urges that we repress. After studying neurology, Freud had the world-changing insight that much of human behaviour appears to be out of a person's conscious control. This was the perfect message for its time. During the nineteenth century, people had become entranced by scientific discoveries that were unveiling hidden worlds. They were learning about genes, bacteria and evolution – unseen forces in the air and in the body that had apparently godlike power over our fates.

For Freud the job of the psychoanalyst, this new form of priest, was to unbury the unseen forces that live within us and bring them into consciousness. 'Freud saw the analyst as an Oedipus figure: a seeker of self-knowledge and knowledge of others, no matter what the cost,' writes Professor Helen Morales. Patients would arrive for psychotherapy with the master at Berggasse 19, in the Alsergrund district of Vienna, to find a consulting room overflowing with books and artefacts from the early years of the Western self. They'd lie back on his famous couch to see, on the wall above them, and just to the right, a framed copy of Jean-August-Dominique Ingres's painting *Oedipus and the Sphinx*. 'To be sure, psychoanalysis was born, in part, from neuroscience, in which Freud was trained,' writes Morales. 'But it was classical mythology that provided the crucial inspiration, scaffolding and legitimation of fundamental psychoanalytic theory.' Without the myths of Ancient Greece, she suggests, 'there would be no psychoanalysis.'

In the fifteenth century, ideas from both Ancient Greece and Rome had undergone a 'renaissance' that would eventually begin pulling the Western self out of its fug. It's not a coincidence, of course, that this renaissance was centred in a place and time that also saw radical changes in how we got along and got ahead. It was in the mighty trading hubs of Genoa, Florence and Venice – glorious centres of hustle and thought that can't help bringing to mind Ancient Greece's 'civilization of cities' – that modern capitalism was born, with its debts, credits, powerful bankers and paper money.

The Renaissance aside, most histories would mark the break between Christianity and the discovery of the unconscious as a time of revolution. This, after all, was the beginning of the age of modern psychology. For all his flaws, Freud was undoubtedly a genius and he certainly prepared vast tracts of essential ground for what came next. But for this particular story, Freud's essential view of the human animal only really constitutes a shift in perspective. Humans were still bad. They still needed to be fixed. The cure remained an eternal war with the inner self, which was morally polluted, purely by dint of birth. Freud was really just a self-hating, sex-afeared, secular reinvention of St Benedict. The actual revolution happened out west, in the United States of America. It was there that our view of who we are, and how we ought to be, underwent a true metamorphosis and gained an armoury of new characteristics, a great many of which we still carry with us today.

In 1936, in an incident that surely seemed to him of little consequence, Freud was visited by an emissary of this revolution. Fritz Perls was a German-Jewish psychoanalyst who'd discovered Freudianism as boy after becoming terrified that masturbation was ruining his memory. After fleeing the anti-Semitism of Europe, he'd found safety and professional success in Johannesburg. Perls was a rather self-important man and, being a disciple of Freud and finding himself in Vienna for a conference, had decided to arrive unannounced in order to pay his respects. He'd presumably expected to be welcomed in and for a great meeting of minds to take place. But when Perls found his master's house, Freud only opened his door a crack.

'I came from South Africa to give a paper and see you,' Perls explained.

There was a silence.

'Well,' said Freud. 'And when are you going back?'

It was an awkward moment. The men had a short, stilted conversation and then Freud closed the door. Perls stalked away, numb with shock and humiliation. He'd never forgive Freud for the snub and would eventually disown him completely. Decades later, and many miles to the west, his new ideas on what a person is and should be would become so influential he'd earn his own place, alongside Freud, in the sack of noisy ghosts we call the modern self.

BOOK FOUR

The Good Self

I'm at my happiest when travelling alone. I love it all: the top deck of a night bus in winter; a window seat in a train carriage that's rocketing past hills and housing estates; the inherent drama of flying: the hustle and thrill of airports, those great flying machines whose noses almost touch the glass walls of the terminal only to be tugged away to soar the planet. I love the cheap food and the forward motion; the sense of mission underway; of crisis, struggle, resolution in play. I love, too, the comfort of the crowd: being enveloped in a mass of others whilst safe in my status as a stranger. They won't talk to me and yet I'll be with them, nourished, somehow, by their proximity. Reading, music, film and miles and miles to go in the company of no one who knows my name: it's the perfect party.

But that flight out of Heathrow was bad. I don't know why. At its worse it would come on like this. I'd feel it physically: a bouldery weight in my throat and upper chest; a treacly rushing in my stomach; a kind of shimmering stinging that could be felt anywhere – the backs of my fingers, the bridge of my nose, underneath my eyes. I call it my iron vest. When it's bad my depression can make me feel separate in some beastly way from all the people around me. But I knew better, by then, than to attempt any explanation of what it all meant. Best to just drift within it, aware it was a part of me that I'd never fully understand and that, at least for now, it had jurisdiction. It was just weather, and some days the weather's bad.

It was night. The shades were down. I turned to the back of my book and made a note on the inside cover: 'We're animals but we think we're not animals. We're products of the mud. We're fooled into believing we're above the behaviour of dogs.'

I closed it, worried that my neighbour might see and conscious that, the next time I read it, I'd probably wince. Hours later, the plane landed. I made it through immigration, located the obscure courtyard for my bus and spent the long journey squashed into the far back corner with my hood up and my earphones in. Eventually, and too soon, we turned left, and there it was, the famous sign: 'Esalen Institute – By Reservation Only'.

The road wound down a steep hill, past lines of huts, and we filed out towards the office, a one-storey wooden structure that was attached to the historied Lodge. A large noticeboard was busy with schedules for yoga, Pilates and meditation, as well as flyers for a 'Straw Building Conference' and something called the 'Feldenkrais Method'. There were appeals for 'Karma Servers in the Lodge' ('Make the bucks. Clear the karma'), new roommates ('We are a wholistically oriented non-toxic household seeking our next fantastic addition') and lifts ('I WOULD LOVE A RIDE TO LA OR ANYWHERE NEAR THERE. HAPPY TO SHARE GAS AND SONGS – LUNA'). On a door, a laminated sign read: 'This is a Door.' Steps ran down to a neat lawn and an outdoor pool, beside which people were being massaged. They looked naked, but I wasn't wearing my glasses and didn't want to squint and risk betraying my competing feelings of intrigue and disquiet. At the end of it all were the cliffs and, then, the ocean, which shone.

*

Remember when the 'geography of thought' expert Professor Richard Nisbett told me, 'the further west you go, the more individualistic, the more delusional about choice, the more the emphasis on self-esteem, the more the emphasis on self-just-about-everything, until it all falls into the Pacific'? This is where it falls into the Pacific. It was on these cliffs that the Esalen Institute helped rewrite our sense of who we are.

Whilst the causes of any specific change in the culture tend to be myriad and are often impossible to isolate, it's clear that much of who we are today has been influenced by what happened here, back in the 1960s. So much of the twenty-first-century self finds a path back to this hundred-and-twenty-acre patch beneath Highway One, from our fetishization of personal authenticity and 'being real' and its concomitant hatred of 'fakeness', to the normalization (not least on social media) of living the intimate details of our private lives as public, to our deep interest in concepts such as 'mindfulness' and 'wellness' – new, secular retellings of Christian narratives of conscience and soul. But, perhaps most portentously, it was also at Esalen that the Western self began being lovingly penetrated by narcissism.

It could only have happened in the USA. For many years America remained a nation under the oppressive grip of the old-world God. It had been settled by Calvinists who, writes Barbara Ehrenreich, lived under 'a system of socially imposed depression' in which 'the task for the living was to constantly examine "the loathsome abominations that live in his bosom" seeking to uproot the sinful thoughts that are a sure sign of damnation.'

But when it finally severed its cord with that old world, it became a radically new kind of place. Its founding document, the Declaration of Independence, stated that 'all men are created equal, that they are endowed by their creator with certain unalienable rights, that among these are life, liberty and the pursuit of happiness.' It would be a land of egalitarianism in which empowered citizens would never cower under the tyrannies of kings and dictators; where they would be free to be who they wanted to be and reach for their dreams. America was exceptional. Its men and women, writes historian Professor Carol George, 'took for granted the reality of a chosen people in a chosen nation, unquestionably marked for social mobility and enduring national greatness'.

These were the starting conditions for what would become

an American revolution of the self that would infect us all. It was a vision, unlike that of Freud or the European Christians, that saw the human as something inherently deserving of good and that contained within it everything it needed to make itself healthy and wealthy and smiling. This was, of course, a partial revival of Ancient Greek ideas of the perfectible human self, with its focus on the all-powerful 'I'. What the two countries shared was their unusual compartmentalized structure: where Greece was a civilization of 'city states', America was a collection of 'united states', their independence from centralized authority, except where explicitly granted, being enshrined in its Bill of Rights. And, suitably enough, it was in this new particulate landscape that individualism would become supercharged.

But, of course, for the people to truly change, there first had to be a change in the ways they got along and got ahead. The long nineteenth century was a time of intellectual and economic revolution in which developments in the realms of science, technology and industry had an explosive impact on the nature of who we were. This was the era of Darwin and Pasteur and Dr John Snow, of steam power, mass production, the railroad, electrification, rising living standards; the beginnings of class mobility; of magnetism, mesmerism, electricity, genes, heredity, adaptation, evolution, germs, infections, forces of nature, all unseen, all-surrounding, inside our bodies, under the earth, out there on the wind. This great storm inevitably stirred the Western self.

In previous eras, when the fate of humans was still so dependent upon the physical environment, that physical environment retained its power to define who we are. But in this new age, when fewer lived on the land, we were increasingly freed from its tyranny. We'd still come into the world asking, 'Who do I have to be to thrive in this place?' but, from now on, it would increasingly be the economy that was the terroir of self, its deep controlling source.

Of course, it wasn't only in America that these changes were taking place. Back in Europe, this economic environment was also creating a new form of ideal self. In 1859, a former journalist, railwayman and political activist (and also, as it weirdly turns out, my great-great-uncle), Samuel Smiles, published *Self Help*, the first book of its kind and a surprise bestseller. His wish in writing it was 'to stimulate youths to apply themselves diligently to right pursuits – sparing neither labour, pains, nor self-denial in prosecuting them – and to rely upon their own efforts in life, rather than depend upon the help or patronage of others.' For someone living in Britain before the Industrial Revolution this would have seemed an impossibly optimistic message. No longer were you obediently stuck in your place – now, with hard work and character, you could improve your lot spectacularly.

This new idea of self spoke of a new world and the new dreams of the individuals living in it. 'Before the eighteenth century power had all been about the landed gentry,' historian Professor Kate Williams told me. 'Smiles was writing in the era of the Industrial Revolution, widespread education and economic opportunities offered by Empire. It was the first time a middle-class man could work hard and do well. But they needed a formidable work ethic to succeed, and that's what Smiles codified in *Self Help*.' The ideal self was now one who relied on neither the lord of the manor nor the Lord of the heavens for sustenance and protection. Although Christianity's heavy presence remained, especially in the deep interest in 'self-denial', temperance and purity of the soul, getting along and getting ahead were increasingly up to the individual, now – and the goal was wealth and improved status. Having staggered out from beneath its religious raincloud, ambition was drying its wings and preparing to soar.

But over in America, at first in the margins, something unique was stirring. Modernity was influencing the Christians in a way that was wholly characteristic of their more optimistic,

I-centred sense of self. In this magical time of invisible forces such as the electric light and the telegraph there developed a craze for faith healing, in which it was thought a believer's maladies could be cured by the touch of a preacher – as long as they had sufficient faith. To be healed of all that ailed them, all they had to do was *believe*. One of its most famous practitioners was an expat Yorkshire plumber named Smith Wigglesworth who once tried to cure a man of stomach cancer by punching him in the stomach. The man flatlined. On another occasion, he kicked a crippled child off the stage. Those who complained that they hadn't been healed were admonished for having insufficient faith. The key to the cure – to happiness, to health, to salvation – lay within them. All they had to do was *believe*.

Adherents of what psychologist William James termed 'the mind-cure movement' thought similarly. James defined mind-cure as the 'intuitive belief in the all-saving power of healthy-minded attitudes'. Mind-cure's forefather was a clockmaker from New England named Phineas Quimby who'd become fascinated by 'magnetic healers' who claimed to have access to quasi-magical powers. But Quimby decided that, actually, their patients only felt better because they believed in the authority of the healer. 'The cure is not in the medicine,' he wrote, 'but in the confidence of the doctor or medium.' He began testing his ideas on the unwell. Writes Mitch Horowitz, 'Quimby's method was to sympathetically sit face-to-face with a patient, never denying that the subject was sick but rather encouraging him to "understand how disease originates in the mind and to fully believe it." If the patient's confidence in this idea was complete, Quimby would then urge the patient to ask: "Why cannot I cure myself?"'

In 1862 Quimby treated Mary Baker Eddy, who went on to found the Christian Science movement, which located the source of all sickness and misery in the mind. Twenty-six years later, a British-born suffragette named Frances Lord, who'd

become entranced with the Christian Science scene on a visit to the US, published what Horowitz describes as 'the prosperity gospel'. Her book was special in that it included 'a "treatment" for overcoming poverty. Lord provided a six-day programme of affirmations and exercises to break down the mental bonds of poverty.' For Samuel Smiles, achievement and wealth had been a product of toil and self-denial. For Lord, all you had to do was *believe*.

Then, in the first half of the twentieth century, came the Great Depression. This utterly catastrophic event was to trigger a series of enormously significant changes to the economy that would alter the selves of the people for decades. This was the first of two crises that would lead to a dramatic reaction against untrammelled individualism. It was the crash, and then World War II, that brought about a new, collective era of 'class compromise' between rich and poor. A series of state interventions led to an extraordinary narrowing of the income gap that economists sometimes call the 'Great Compression', a period that lasted roughly between 1945 and 1975. And, sure enough, out of this new kind of economy hatched a new kind of self.

It all began with the 'New Deal' of the 1930s, which brought tough regulations on banking. Then there was the Social Security Act and the introduction of the minimum wage. Unionization became popular. The tax level for the nation's highest earners was set as high as 90 per cent. The GI Bill gave millions of working-class war veterans a college education paid for by the state, a big-government act that would do immeasurable good. It was partly due to policies such as these that, between 1929 and 1945, lower incomes would grow faster than higher and, for the next twenty-five years, wages grew at about the same rate for everyone, rich or poor. This Great Compression led to 'a golden age for millions of high school graduates', writes the economist Professor Robert Gordon, 'who without a college education could work steadily at a unionized job and make a high enough income to afford a suburban house with a backyard, one or two cars and a life style of which median-income earners in most other countries could only dream.'

Encouraging this new collective spirit was the rise of industrial automation. Farm workers moved in huge numbers to the towns and cities. There, they no longer lived side-by-side with relatives and old acquaintances but with strangers, on whom you had to make a good impression. The post-war years were a time of salesman and corporation, of the company taking care of you for life. Increasingly the individual became just a small component of the larger corporate organism. And people were living that way, too, in the growing, bustling, faceless suburbs. To get along and get ahead in this new kind of tribe, you had to be accepted by the group. That meant being witty, smart, open, upbeat and attractive. The emphasis on hardiness of character transformed into an emphasis on maintaining a sunny and winning disposition.

It was around this time, as Susan Cain has famously documented, that the provost of Harvard University began instructing admissions officers to reject applications from 'sensitive' applicants and instead offer places to 'healthy, extroverted' ones. There were new runaway bestsellers, American retellings of the Samuel Smiles narrative that had arteries of Mind Cure and faith healing running deep within them: *How to Win Friends and Influence People* by Dale Carnegie ('Men and women can banish worry, fear and various kinds of illnesses, and can transform their lives by changing their thoughts. I know! I know!! I know!!!') and *The Power of Positive Thinking* by Reverend Dr Norman Vincent Peale ('If we are our own chief problem, the basic reason must be found in the type of thoughts which habitually occupy and direct our minds'). They contained such lessons as 'how to make people like you instantly', 'how to create your own happiness' and how to 'expect the best and get it'. As sociologist Professor John Hewitt so adroitly puts it, 'The era of "character" vanished and that of "personality" arrived.'

But this new collective mood, so alien to the individualist American core, didn't come without costs. Nightmares wafted in from the disturbed national subconscious. These were the years of Red Terror and McCarthyism in which the paranoid idea spread that America might tip into Communism. There was a similar feeling of rising dread over the coming computer age. It was feared the future would be a 'technocracy' in which freedom and individuality would be crushed and the population dominated by machines of coercion, conformity and control. Computers were seen as a war technology that would be co-opted by the collective powers of government and corporation and used against us. 'The dominant minority will create a uniform, all-enveloping, super-planetary structure, designed for automatic operation,' wrote Lewis Mumford in his 1967 book, *The Myth of the Machine*. 'Instead of functioning actively as an autonomous personality, man will become a passive, purposeless, machine-conditioned animal whose proper functions, as technicians now interpret man's role, will either be fed into the machine or strictly limited and controlled for the benefit of de-personalised, collective organisations.'

As the 1950s became the 1960s, the Great Compression created a new and distinct generation of selves. The children of Corporation Man and Woman were the even more collectively-minded hippies who changed the Western world with their ideas of community, anti-authoritarianism, anti-capitalism, pacifism and imported Eastern notions, which suddenly made much more intuitive sense, about the general oneness and connectedness of all things. This state of mind also saw a major shift in the attentions of the political left, away from their traditional preoccupations, such as employment conditions and pay, and towards equality and rights for minorities.

The American transformation of the Western self would be a product of the Great Compression. It combined Corporation Man's sunny, extrovert bearing with Mind Cure's belief in the power of thought with the nation's core sense of optimism and exceptionalism. From that wild recipe came a new form of individual and a new form of individualism. It would give birth to an era in which enormous expectations were placed on the self, which, it was now believed, was filled with vast reserves of incredible yet hidden potential. This revolution was led by a cultural leader who might not be as famous as Aristotle, Jesus or Freud, but who's managed to have an outsized influence on us all. Any story of who we are today would be absurd without his presence. His name is Carl Rogers.

During his early life, Rogers retraced the journey of the Western self in microcosm: he began as a personally ambitious boy who believed in the power of education to better himself, which was all very Greek, but, on top of that, he was also a devout Christian. That changed when he made a five-month trip to China as a theology student, and witnessed horrors – including child labourers in a silk factory and the inside of a prison – that for ever changed him. 'As a result of seeing all this human suffering, Carl lost his faith,' his biographer, David Russell, told me. 'When he got back to university, he walked across the street from Divinity to Education and got his Ph.D. in psychology and psychological education.'

During the 1930s, Rogers was working at the Child Study Department of the Rochester Society for the Prevention of Cruelty to Children. In a direct echo of the revelation that struck Phineas Quimby almost a century earlier, he fell upon the notion that there was something in the relationship between therapist and patient that seemed to play some critical factor in their recovery. They should be treated, he came to believe, not as if they were somehow sick or dirty, but in a characteristically American spirit of optimism and trust. The sick had the resources within them to heal but, to do so, they needed to be freed from the critical judgements of the shrink and society. They should be received with an attitude that he christened 'unconditional positive regard'. The thinkers of old

Europe had believed that people were bad and needed to be controlled. For Rogers, these authoritarian dinosaurs could hardly have been more wrong. On the contrary, he wrote, 'the innermost core of man's nature, the deepest layers of his personality, the base of his "animal nature", is positive in nature – is basically socialized, forward-moving, rational and realistic.' Just like Aristotle, he became convinced that humans naturally moved towards perfection. But in order to be happy, they needed to be set free of society's judgement and disdain.

As an eighty-three-year-old, in 1985, Carl Rogers was still complaining that Freud, that 'authoritarian European', was being taken seriously. Freud's ideas of psychoanalysis were 'seductive and disgusting', he told his biographer, and it was 'socially reprehensible' that people hadn't moved on. Even the Freudian tradition of the therapist sitting out of view behind the patient who lay on the couch was 'somewhat repugnant'. He apparently wasn't joking when he recommended that a study should have been made of the children of Freudian analysts. 'If one needs any evidence as to the failure of the psychoanalytic point of view, it certainly resided in the children of psychoanalysts. They were a mess, almost without exception.'

In the 1960s, Rogers helped found the discipline known as Humanistic Psychology, which, in turn, inspired the Human Potential Movement, whose adherents believed in the incredible power of the individual and our almost unlimited capacity to transform into better and better versions of ourselves. Rogers and his disciples became convinced that the non-sick could benefit from psychological work. 'We had the idea that if it was good for neurotics, it would be good for normals,' Dr William Coulson, chief of staff at Rogers' Western Behavioral Sciences Institute in California, has said. Rogers created 'encounter groups', therapeutic spaces in which people, under his guidance, were freed of the usual social expectations and permitted to be honest with themselves and others, thus creating an atmosphere of trust, daring and 'radical authenticity' in which they could burrow through to each other's perfect core, leading to breakthrough and transformation.

In 1964, Rogers received funding for a three-year study that would develop the encounter idea of using a network of institutions run by the nuns of the Immaculate Heart Community in California. In a 1994 interview, Coulson, who took part in the study, described it as 'a disaster'. They began by showing the nuns a film of an encounter group. 'The people in that film seemed to be better people at the end of the session than they were when they began,' he said. 'They were more open with one another, they were less deceitful, they didn't hide their judgments from one another; if they didn't like one another they were inclined to say so; and if they were attracted to one another they were inclined to say that too. So they went along with us, and they trusted us.' But, according to Coulson, this new-found authenticity unleashed a firestorm of lesbianism and rebellion. 'There were some 615 nuns when we began. Within a year after our first interventions, 300 of them were petitioning Rome to get out of their vows. They did not want to be under anyone's authority, except the authority of their imperial inner selves.' Rogers cancelled the study a year early. 'We thought we could make the IHMs better than they were,' Coulson said, 'and we destroyed them.'

But encounter would only become more intense, more dangerous – and more popular. Four years before the start of the nun experiments, another of the Human Potential Movement's pioneers, the author Aldous Huxley, gave a lecture at the University of California, Berkeley. In it he argued that the incredible and rapid changes that had taken place over the recent decades were a demonstration of the awesome potential that lay latent in humans. He marvelled at our future possibilities. After all, he said, in a confident yet utterly false assertion that's still believed by many today, 'neurologists have shown us that no human being has ever made use of as much as 10 per cent of all the neurons in his brain. And perhaps if we set about it in the right way, we might be able to produce extraordinary things out of this strange piece of work that a man is.'

Huxley advocated the foundation of a kind of base camp for exploratory research into the 'human potentialities'. Sitting in the audience that day happened to be a man who'd had a strikingly similar idea. Twenty-nine-year-old Richard Price would, along with his partner, fellow Stanford psychology graduate Michael Murphy, go on to establish the Human Potential Movement's most holy site. Both Murphy and Price were good sons of America's collective economy, having keen interests in Eastern spirituality, with Murphy having spent eighteen months in Puducherry, India, practising meditation.

What began as a place for earnest lectures became much weirder, in 1963, with the arrival of Carl Rogers-inspired encounter groups, which were described by Murphy as 'intense stress sessions, often lasting at least forty-eight straight hours, in which fifteen or more participants meet in one room to develop and discuss their feelings for one another.' His and Price's 'base camp' was to become a place of legend, scandal and suicide; a place where Rogers, along with most, if not all, of humanistic psychology's most iconic thinkers and leaders held workshops and increasingly crazed sessions. It was a place that rang with ideas that would change America's self and then the world's. It was a place called Esalen.

The Esalen Institute's stated mission was that 'all men somehow possess a divine potentiality; that ways may be worked out – specific, systematic ways – to help, not the few, but the many toward a vastly expanded capacity to learn, to love, to feel deeply, to create'. Building on Rogers' idea of 'therapy for normals', they believed that God lay in the deep, authentic core of everyone and what we had to do, using methods including encounter, was expose it. Sociologist Professor Marion Goldman writes that one of the reasons Esalen attracted thousands of Americans was because it 'fundamentally redefined psychotherapy as a context for personal growth rather than recovery from mental illness'. As well as therapy, it was a pioneering popularizer of the kind of 'spiritual' practices, such as yoga, massage and meditation, that were concerned with veneration of the god of the self rather than any found in a holy book. 'Esalen played a critical role in introducing and promoting esoteric spirituality so that it flowed into mainstream culture,' writes Goldman. 'Millions of contemporary Americans identify themselves as spiritual, not religious, because the Institute paved the way for them to explore spirituality without affiliating with established denominations . . . The basic assumption that God is part of all beings and that we are gods is Esalen's cornerstone.' This was a specific form of spirituality that placed the source of divine perfection within the self. It was an idea that would eventually swallow the world.

*

Eight thirty p.m., map in hand, having plotted my path amongst buildings named Maslow, Huxley and Fritz, I finally arrived at the Big Yurt. It was a small round structure, with wooden floors and ceiling, that had been set up like a theatre. There was a stage and powerful lights that beamed towards its centre. I sat myself nervously on the end of the back row, one of about twenty others, and watched Paula Shaw, the course leader, enter the spotlight. I knew almost nothing about what was going to happen, apart from that it would be ongoing for six days and we'd start in the morning and work late, often until midnight. Months earlier, the woman on Esalen's booking line had told me that this was the closest thing they had to the encounter groups that a visitor might have experienced in the 1960s. The catalogue explained that it held 'honored status' at the Institute and was considered a 'rite of passage' for staffers. It would be a 'voyage through your own humanity' from which we'd emerge with 'greater authenticity'. It was, went the promise, an 'opportunity to experience yourself in ways you may have dreamed about but never thought possible'. It was called 'The Max'.

Paula Shaw was on a seat at the centre of the stage, scanning us silently, teeth sitting on lower lip, eyes filled with menace and delight. She was born in the Bronx in 1941 and, it would quickly become clear, remained as tough and sharp as any Bronxite, even if, at the age of seventy-four, a little deaf. She said nothing. She continued saying nothing. The longer her silence continued, the greater became the unease. People began shuffling, coughing, tittering. And still she said nothing. It was masterful. Paula had us, totally. Then, without warning . . . 'WELCOME TO THE MAX!'

With the tension broken, and a brief welcome made, she began issuing orders. 'You're now in process,' she barked. 'Once you've agreed to be in, that's it. You're in. There's no leaving. I don't want you talking to people outside the workshop. When you have a break, no walking off. You don't go back to your rooms. Confidentiality. Feel free to share what you see in here, but don't attach names. And don't bring your bottles of water. People these days, always sucking on their water.' She mimicked a drinker slurping at a plastic teat. 'And, please, this week, stay off the internet. Do not use it. You're *in process*. And don't talk to each other about what's going on in here. Don't process the process. If people get upset, if there are tears, don't comfort them. What do we do when we hug?' With an oleaginous smile, she mimicked someone hugging. 'We smother it. Comforting is about making *you* feel better.'

Falling quiet once more she slowly took in all of our faces and began nodding, as if coming to some secret conclusion. 'This is not one of those *pillow* courses,' she said, a brief sneer flashing across her narrow lips. 'This is about expanding who you are. I've been doing this over thirty years and, I'm telling you, it's heavy stuff. We're about transforming people's lives. OK? So here are the basics. There are three rules, and the first one goes back to Aristotle: a thing is what it is.' She stood and lifted her seat in front of her. 'This is a chair. It has a plastic back and metal legs. It has certain properties, weight, height, depth. It is what it is. Got it? OK. Rule number two: own it. Rule number three: be creative with it.' She sat back down and grinned. 'I hope you're ready. Because we're going to light a firecracker under you.'

As Paula had been speaking, an atmosphere of mewling, pawing emotionality had opened up in the audience. The air was slippery with it. Attendees in the front row were staring up at her, eyelids stretched as if trying to reach towards her with their very pleading globes. They sat with steepled hands pressed to their hearts, laughing in all the right places. I watched from the back, trying to shrink myself into my chair.

We were told that, one by one, we were to stand on the stage and find someone in the audience to 'anchor' ourselves to by meeting their gaze. Then we were to describe the sensations we were feeling in our bodies. We weren't allowed to label these sensations – we couldn't say we were 'nervous' or 'nauseous' or whatever – because this was the language of our 'system'. We were here to rediscover the reality of what we actually felt by describing the feelings themselves rather than the labels we'd learned to put on them. The aim was to be like children who naturally express themselves freely before they're beaten down by adult society. Children 'are who they are', Paula has said. 'The thought of going outside your comfort zone, to be able to be all that you can be, is scary. It's threatening to the way most of us live our lives, to the box we live in.'

I crouched forward in my seat, curious as to what would happen next. This reality versus the 'system' idea, it reminded me of the left-brain interpreter and its confabulations. We feel what we feel, and the voice in our head tries to label and explain these feelings, even though it has no direct knowledge of them. Maybe Paula was onto something. When she'd said that the course's first rule 'goes back to Aristotle' I was briefly electrified. *A thing is what it is. Own it.* No matter how much I might have wished I was back in England in bed with my dogs, the moment she said that, I knew that I was, once again, in exactly the right place.

Up beneath the spotlight a pattern quickly emerged. A person would step onto the stage and stare into the eyes of another before describing their physical feelings. They would then start talking about their fear or sense of inadequacy or shame or whatever. Then, Paula would say something like, 'Someone in your past has made you feel that way. Who was it?' And almost every single time the answer was 'my father'. Paula would tell them to 'put him out here' – to imagine the person in the audience that they were 'anchored' to was their dad. 'What do you want to say to him?' And then tears would run.

There was a woman in her sixties with long grey hair, a deep tan and bloodhound eyes who said she felt a shaking in her legs. 'Breathe into the shaking,' Paula said. So she did. The shaking became visible. Paula asked her, who was it who made her feel like this? 'My father,' she said. 'So put him out there.' The woman's arms began shaking and then her shoulders. She was now sobbing at the imagination of her dad, berating him madly. 'You wanted me to be an engineer and think in a mathematical way,' she wept. 'I asked for a chemistry set just to please you.' The shaking was getting worse and worse. Her eyes were red and streaming. Her voice rose an octave as she wailed into the blackness: 'You always wanted me to think linearly.' By now her entire body was vibrating. She was bouncing up and down. Paula asked, 'Is there a sound that connects with the shaking?' Jumping up and down, over and over and over, inches off the stage, she began shouting: 'Huk! Huk! Huk! Huk!'

And so it went. One woman had a father who'd overtaken her in a running race and, as a result, she'd grown up to be too competitive. Another had a father who'd made her feel 'invisible' and now, as an adult, that's how she lived her entire life. Another tearfully complained that her dad once 'called me a cunt as if it was nothing'. And then one of the workshop's handful of men climbed onto the stage. He was a handsome, broad-shouldered Finnish architect whose father, a successful musician, had wanted him to be a famous cellist. 'I remember being seven and you watched me play with a face that said, "I can't believe how much you suck,"' he said, weeping at the memory. 'I feel so angry. I feel it here.' With a squeeze and a grimace, he grabbed his cock and balls.

Having learned what I had about male suicides, and the impossibility many men feel at achieving the perfect model of caring but powerful manhood, I found myself feeling rather sorry for all these hated fathers. I shifted on my seat, my fingers flexing, my leg jiggling, as attendees came up on stage and then left. I counted them down, one by one by one. As my own inevitable turn beneath the spotlight drew nearer, I began to want, very badly, to disappear.

*

Fritz. That's what you called him, despite his celebrity. Just Fritz. But don't mistake the familiarity for friendliness. Fritz would treat you with ferocious contempt and he would do it in front of everyone. He'd been raised tough, by a mother who'd punish him with whips and carpet-beaters and a father who'd call him a 'piece of shit'. He was expelled from school, rejected and humiliated by a prostitute at thirteen years old, fired from an apprenticeship and then, finally, driven by his terror at the deleterious effects of his masturbation, he'd trained as a psychoanalyst. He left Germany in 1933, after being placed on a Nazi blacklist, and eventually settled in South Africa. There, he found success and wealth as a psychoanalyst and he remained a loyal, if revisionist, follower of its credo until his crushing encounter in Vienna with the master himself, 'a painful and in some ways historically significant meeting', writes Esalen's brilliantly entertaining biographer, Walter Truett Anderson, for which 'he never forgave Freud'.

Around the Institute, Fritz became known for his near obsessional attacks on his former guru, insulting him (as well as members of the Esalen staff) at every possible opportunity. For him, the perfect self was utterly authentic, never censoring itself nor compromising its behaviour. Rather than repress his sexuality, as in the old European style, Fritz would broadcast it to all, his erection arriving before him as he strode naked around its famous hippyish hot spring baths. He'd make passes at nearly every female who happened into his orbit, stroking the genitals of any he found acquiescent. He took to wearing slippers and a onesie and would approach beautiful young women with the radically authentic boast that he was a 'dirty old man', a line that, on at least one occasion, was met with the dream response, 'And I'm a dirty young girl,' and what followed, apparently, followed.

Fritz was known for his 'Gestalt' encounter groups, elements of which I'll be putting myself through at the Max. He rejected Freud's model of rooting through the past for the answers to your problems and, instead, pushed you to confront the absolute truth of you in the now, no matter how painful or difficult that truth turned out to be. His groups operated 'Strictly on the "I and thou, here and now" basis,' he said, warning that 'any escape into the future or to the past is to be examined as a likely resistance against the ongoing encounter.' He'd invite you to sit on what he called the 'hot seat', where you might be asked to describe the physical feelings in your body – sweaty palms or tingling fingers – or to confront your mother or father, who you'd imagine sitting opposite, or to explore an inner conflict by playing two different aspects of yourself, giving voice to the arguments of each.

The idea was to hurl yourself into each role, without self-consciousness, really experiencing it and, as you did, Fritz would remorselessly pick out all the ways in which he believed you were being 'fake', from your tactics of self-serving diversion to the way your eyes moved or little finger twitched. Body language was of particular fascination to him – 'I disregard most of what the patient says and concentrate on the nonverbal level as this is the one that is less subject to self-deception,' he said. If you didn't bore him too much, in which case he might completely ignore you or fall asleep, you'd likely be called a 'weeper', a 'shithead' or a 'mindfucker'. During one 1966 session, Natalie Wood – star of *Rebel Without a Cause* and *West Side Story* – was accused by Fritz of 'absolute phoniness' and told she was 'nothing but a little spoiled brat who always wants her own way'. He grabbed her and, against her will, put her over his knee and spanked her.

Attendees would be expected to 'own' the truth of what their outward behaviour was apparently betraying about their inner self and take responsibility for it. If you found the treatment difficult and wept, then you'd be mocked. If someone went to the assistance of the crying person, they too would be mocked. His goal, he said, was to 'manipulate and frustrate the person in such a way that he's confronting himself'. He was aiming for a state of radical authenticity, an unapologetic embracing of the true, core self; 'to change paper people into real people'. A person's principal goal, he believed, should be to 'be what you are'. These are ideas, of course, that are still very much with us, not least in the realms of social media and reality television, where exposing the details of your private life and being 'real', no matter how ugly it might make you seem, are often encouraged.

The mid-1960s were incendiary and magnificent times for Esalen. Their 1965 brochure boasted that 'new tools and techniques of the human potentiality – generally unknown to the public and to much of the intellectual community – are already at hand; many more are presently under development. We stand on an exhilarating and dangerous frontier – and must answer anew the old questions: "What are the limits of human ability, the boundaries of the human experience? What does it mean to be a human being?"' In 1966 alone, around four thousand inner-adventurers – 'doctors, social workers, clinical psychologists, teachers, students, business executives, engineers, housewives', according to a contemporary *New York Times* account – made the long drive up Highway One to find out. A phrase heard often around the property was 'Mother Esalen gives permission' and, indeed, she did. There was nude massage, group sex, physical aggression and colossal psychedelia; there were grown men re-experiencing their births, young women in sheer gowns playing flutes by the pool and encounters with visionaries such as Ken Kesey, Joseph Campbell and Timothy Leary. Fritz had his wall decorated with the abandoned spectacles of clients he claimed had regained perfect sight following his sessions, while in another group, three women were said to have experienced spontaneous orgasm. Jane Fonda came to Esalen to learn Zen and had an affair with co-founder Richard Price while she was at it. The other co-founder, Michael Murphy, described what was taking place at Esalen as nothing less than a 'consciousness revolution'.

Despite the fractiousness of Fritz's relationship with the Institute's founders, they built him a house overlooking the hot spring baths, at the enormous cost of $10,000. He was even given his own section in the brochure. But the King of Esalen was soon to find a rival. Will Schutz was three decades his junior, had taught at Harvard and the Albert Einstein College of Medicine and had just written a book that, in his first months at the Institute, would grow into a national bestseller. Its title was *Joy*, a state Schutz described as 'the feeling that comes from the fulfilment of one's potential'. Inspired by Carl Rogers, Schutz believed that humans are born with all the joy they'll ever need inside them, but society gets in the way of it, suppressing it as we suppress our true selves.

With this came an extreme position on self-responsibility

that overturned the old Christian model, in which God had a plan for every one of us, and all we could do to influence it was be good and pray for mercy. Now that God was to be found in the self, rather than the heavens, it naturally followed that the source of all our fortunes was similarly relocated. For Schutz, everything that happened to us, including illness and accidents, was invited by the sufferer. 'There is no such thing as a victim of circumstance,' he said. 'The cause of every illness or injury is within, and only the patient can heal himself. You are the one who chooses to be ill in the first place.'

Schutz's encounter groups typically involved an individual being pushed into a state of radical truth not by the therapist, as in Fritz's Gestalt model, but by their fellow group members. But like Fritz, he preached that in order to be free and joyful, people needed to be 'real'. Only by knowing the truth of themselves could they be set free. One of his followers put it this way: 'Under this Encounter Contract I say how I feel about you. My obligation to be polite, kind, or considerate is, for the time, set aside. This Encounter Contract replaces the familiar Social Contract.'

His groups could be brutal and strange. In one process, named Pandora's Box, women would work through their negative feelings about their vaginas by exposing them to the group. Husbands and wives were told to pick three secrets that would destroy their marriage and confess them all, an experiment which lead to at least one woman bloodily assaulting her partner. A man who complained of feeling 'held down' by life was buried beneath a pile of humans until he screamed and was forced to fight his way out. An introverted psychology teacher named Art Rogers was assailed by his group for 'intellectualizing' the process and being insufficiently 'real'. A woman grabbed his pipe and smashed it. He was chased away by the men and threatened with a beating. When he returned, that afternoon, they began taunting him, as if in a playground nightmare, '*Art! Art! Art!*', and he snapped, throwing an assistant through a window, which shattered. The session was considered a success. And if Art happened to feel in any way damaged by the experience, then that was up to him. 'Like Fritz,' writes Anderson, 'Schutz made the taking of responsibility for a client sound like a betrayal of the human potential.'

Fritz and Schutz maintained a peace, for a while, despite their pitifully masculine interest in the size of each other's encounter groups. But the huge national success of the Institute began drawing the big media – and they seemed interested, mostly, in Schutz. Surprisingly, a September 1967 article in *Time* didn't mention Fritz or Gestalt at all. Fritz began sniping. He disparaged Schutz's clients as mere 'joy boys' and would publicly insult him in the Lodge, commenting acidly that everyone in his groups seemed to be having fun. Next, a writer from the *New York Times Magazine* came to Esalen – to take part in a Schutz workshop. Then a *Life* magazine writer came to do the same. Schutz was a guest on *The Tonight Show* starring Johnny Carson three nights in a row. In a move that must have infuriated Fritz, he named himself 'the first Emperor of Esalen'. But the feud that was escalating between the giants was just one of the wounds that would ultimately bleed Esalen's glory phase to its death.

Next came the suicides.

*

The authentic self is godlike. Our true thoughts and feelings shouldn't be repressed behind the old-fashioned curtain of 'manners'. We should 'be real' and disparage those who are 'fake'. Our self is its own justification. What the self wants, it should have. What the self thinks, it should say. *The innermost core of man's nature, the deepest layers of his personality, the base of his 'animal nature', is positive in nature.*

These ideas, which seemed so radical in the Big Sur of the 1960s, have grown to become enormously prevalent today. The problem is they're based on a premise that's false. What

Carl Rogers and the intronauts of Esalen couldn't know is that many of today's experts claim there *is* no authentic self. Rather than there being a pure and godlike centre to us all, we actually contain a collection of bickering and competing selves, some of whom, as we'll see, are quite disgusting. Different versions of 'us' become dominant in different environments. It's now often claimed the human self cannot be reduced to some 'innermost core'. The 'I' is not one, it is many.

One sunny spring morning, I travelled to the office of Professor Bruce Hood, the famous developmental psychologist at the University of Bristol. I wanted to find out what he means when he calls the self 'a powerful deception generated by our own brains for our benefit'. He motioned me to sit in the small area of the room which comprised various seats around a low coffee table. On a bookshelf to my right was a hand holding a grenade, a human skull and a photo of a Victorian man glaring ferociously from behind a splendid beard. A whiteboard with the words 'Meet Your Brain' hung on the wall behind him.

'At its very simplest, a self is a way that we can make sense of the things that happen to us,' he told me, leaning back in his chair with his legs crossed out in front of him. 'You need to have a sense of self in order to organize your life events into a meaningful story.' The notion that you have a soul, he explained, is an illusion. We feel as if we have a magical centre, a special core in which our moment-by-moment life is experienced. But there is no centre. There is no core. There is no soul. We experience our thoughts as if we're somehow listening to them, but as the philosopher Julian Baggini has observed, 'our minds are just one perception or thought after another, one piled on another. You, the person, is not separate from these thoughts.'

According to Bruce, the illusion that we have a stable 'authentic' self begins with us looking out at the world and other people, and seeing how they treat us. That's how we build our model of who we are. This idea is sometimes referred to as 'The Looking-Glass Self'. In his book *The Self Illusion*, Bruce cites its creator, the sociologist Charles Horton Cooley, who wrote, 'I am not what I think I am and I am not what you think I am; I am what I think that you think I am.' The illusion is thought to take form at around the age of two. 'That's when you start to have autobiographical memories,' he told me. 'Then, in most cases, at around two to three years of age, children start interacting with other children, competing with them and joining groups. To participate usefully in a social arrangement, you have to have a sense of who you are, you have to have a sense of identity, and that sense of identity is a self. That must be constructed by the contextual information – Who am I? Which groups do I belong to? – as well as the biological ones – Am I a boy or a girl? Am I black or white? – or whatever. You then merge these and form in-groups. You start to develop prejudices and biases. You become pre-occupied with what others think about you. Your sense of self worth is a reflection of what you think other people think about you. As you spend more and more time with other children, you start to develop hierarchies.'

With these hierarchies comes our obsessional concern with status. 'We constantly seek validation. Why do we buy the fast cars? Why do we own the big yachts? Why do we need all these things that we don't really need? It's because we're signalling to others our self-status.' If others believe we're fancy and great, our looking-glass self interprets that as evidence that we *are* fancy and great. We're self-conscious. We use clues from out there to tell us who we are in here.

And that, until the end of our days, is pretty much how it goes. We dread insult and crave good reputation. 'The rest of your life is, I think, motivated by feeling good about ourselves and avoiding negative emotions, which are about being ostracized or rejected or devalued,' he said. 'If you look at our daily activities, we earn wages to bring an income. But once we've satisfied our basic food, housing and environmental needs, we're motivated to pursue validation from other people. And that, of course, is because of this need to keep our belief in our self-worth.'

Our lack of true authenticity means that who we are and how we behave tends to shift, somewhat, depending on where we are and who we're with. When we're doing our job, for example, we often *become* that job. We start to behave as we think the model writer or the model hedge-fund manager or the model teacher would act, aping their mannerisms, dress and code of ethics. Perhaps the most famous chronicler of this effect was the philosopher Jean-Paul Sartre, who noticed it in a waiter, who he accused of acting in 'bad faith': 'His movement is quick and forward, a little too precise, a little too rapid,' he wrote. 'He comes toward the patrons with a step a little too quick. He bends forward a little too eagerly; his voice, his eyes express an interest a little too solicitous for the order of the customer. Finally there he returns, trying to imitate in his walk the inflexible stiffness of some kind of automaton while carrying his tray with the recklessness of a tight-rope-walker.'

Sartre was observing an individual who seemed to have been taken over, possessed, by a concept that existed outside him called the Waiter. This man's own voice, body language and movements had been subsumed in service of a cultural idea, this model of the perfect bringer of coffee and cakes. Of course, lots of us do this. 'There is the dance of the grocer, of the tailor, of the auctioneer,' Sartre wrote. In a sense, he argued, this is what society demands of us. 'A grocer who dreams is offensive to the buyer, because such a grocer is not wholly a grocer.'

Sartre's thoughts on the waiter were published in 1943, but today's social psychologists are still interested in the notion that the self warps and morphs depending upon what it believes is expected of it. We have a self for work and a self for home, a self for lovely restaurants and a self for roadside diners; a self for Twitter and a self for Facebook; a self for the plumber and a self for the mayor; a morning self and an evening self, a

Monday self and a Sunday self, a self of the business suit and a self of the dressing gown. The gangster John Pridmore reflected, with befuddlement, on how he could weep when at home watching *Little House on the Prairie* and then go to work and quite happily 'beat someone senseless'. Many of us complain that when we visit our parents for family events, such as Christmas, we seem to helplessly revert to our childhood selves. This is likely because Mum and Dad are treating us as the person we were.

As we travel through our days and lives, then, we're being continually changed by the situations we're in and the personalities that orbit us. The people around us create a kind of psychic mould that we expand into. 'This notion that we have a coherent self with integrity is slightly undermined by the fact that under different circumstances in different events we behave in totally different ways,' Bruce told me. 'These different selves reflect the fact that these are different social environments that we're occupying.'

A team led by psychologist Professor Mark Snyder examined some of these effects in a fascinating study that looked at how physical beauty changes people's behaviour – and how that change in behaviour then changes us. Fifty-one men spoke to fifty-one women via an intercom system. Each man was given a Polaroid of a woman and told (falsely) that it was of the person they'd be chatting to. Some of the Polaroids were of notably attractive women and others were not. Analysis of the conversations that followed indicated that the men appeared to engage with the women under the power of the cultural belief that 'beautiful people are good people' – that old Greek idea of *kalokagathia*. Even though the photos were fake, when the conversations were finished, the men perceived the 'pretty' women as being more friendly, likeable and sociable. But the really fascinating thing about this study was that a significant number of the 'pretty' women *actually began to behave in a way that was more friendly, likeable and sociable.* According to the researchers, not only did the men 'fashion their images of their discussion partners on the basis of stereotyped intuitions about beauty and goodness of character, but these impressions initiated a chain of events that resulted in the behavioural confirmation of these initially erroneous influences.' Other researchers have found that when people falsely believe they're conversing with someone lonely, they tend to behave less sociably towards them and with greater levels of hostility – which, of course, changes the behaviour of the lonely person for the worse, creating an unhappy feedback loop.

Although we obviously have *some* awareness of our shifting guises and behaviour, we're only usually conscious of them when another person points them out, or when the shifts are so flagrant they can't be ignored. However, we're mostly unaware of the constant state of flux that our self is in. It's not even as if we ordinarily have any proper control over it. According to Bruce, it's the environments that really do the switching, not us. And it's mostly unconscious.

And it's not only the *social* environment that changes us. The psychologists Dan Ariely and George Loewenstein explored our multiple nature in an unforgettably dark experiment on twenty-five male Berkeley undergraduates. They were asked to predict how they'd behave in a series of immoral, unusual or extreme sexual circumstances. Would they always use a condom? Could they imagine becoming sexually excited by contact with an animal? What about being attracted to a twelve-year-old girl? Would they slip a woman a drug to increase the chance she'd have sex with them? On one occasion, they were simply were required to answer the questions. But on another, they required to do so whilst masturbating to pornography and in the peak state of 'sub-orgasmic arousal'. The results were disturbing. When aroused, their predictions of whether or not they'd engage in the unusual practices, such as intimacy with animals, were nearly twice as high. Their imagined willingness to act immorally in the pursuit of sex had more than doubled.

The study didn't only suggest we have radically different moral codes depending on our shifting states of self. More unsettlingly, it showed how poor we can be at predicting our own behaviour. We're not one person, then, but many, and the people we are can be strangers to each other.

The unsettling fact of our multiple nature holds, too, on the neurological level. The neuroscientist Professor David Eagleman writes that our brains are 'built of multiple overlapping experts who weigh in and compete over different choices.' One of these 'sub-agents' might argue for sex with a sheep, say, whilst another might argue against. These experts are constantly battling, trying to get us to do this or that, and how we eventually behave depends on which module wins the fight in any given moment. All of this, needless to say, happens beneath our conscious awareness, whilst the left-brain interpreter narrates all our actions, reassuring us that *we* are the free choosers of our behaviour, when we're probably not.

What all this work suggests is that a foundational idea of the Humanistic Psychologists is simply wrong – there *is* no authentic core to us, no essential, happy and perfect version of the self that can be exposed by stripping back the repressing expectations of society. In fact, the self is modular. We're made up of many competing selves, all of which are equally 'us', and which fight for dominance. Different versions of us emerge depending on where we are, what we're doing, who we're with and how aroused we happen to be. Our sense of who we *actually* are turns out to be critically dependent on what we believe others think of us. This brings to mind what Rory at the Suicide Lab told me about social perfectionism: '*It's nothing to do with what people actually think of you. It's what you think other people expect.*' The idea of the Looking-Glass Self exposes the social perfectionist in every one of us. We *all* judge ourselves by looking into the eyes of other people and imagining what they're thinking.

But it also exposes a terrible danger in the concept of the

encounter group. If a person – especially a psychologically vulnerable person – is surrounded by others who are abusing them, perhaps calling them 'weeper', 'shithead', 'mindfucker' or 'fake', then they'll be susceptible to believing that *that's who they actually are.* It also exposes the inherent cruelty in the hard form of personal responsibility that gained traction at Esalen and remains common today. The humanistic psychologists were convinced the authentic self was not only real but perfect and full of latent potential – *only 10 per cent of your brain is in use!* But if it's true that we hold within us all the power we need to succeed, then it naturally follows that if we fail then it's our fault and our fault alone. This is a cruel and narrow line of thinking that runs backwards from today's age of perfectionism, through Esalen's Will Schutz and Fritz Perls, and directly back again to the faith healers and Mind Cure proponents of nineteenth-century America such as Smith Wigglesworth and Mary Baker Eddy. You just didn't want it badly enough. You just didn't *believe.*

*

All I could see behind the lights of the stage were rows of faces in silhouette. From somewhere off to the left, the voice of Paula. 'Tell me what you feel.' The plan was to get all this over with as quickly as possible. I'd be the grey man: calm, efficient, translucent, forgettable. I'd give her nothing. I'd fake it. There was no other way.

'I've got sweaty hands,' I said.

'No,' she barked. 'First you need a pair of eyes to anchor yourself to.' I squinted into the blackness. 'I'd like to point out how thoroughly your system has you,' said Paula. 'Even though you sat there and watched I don't know how many people going up on the stage, you just jumped up, thinking, "Oh, I know what to do." So we see your limited listening. But that's your system. Its protection. Its defence.'

'Yes,' I said, with a level of irritation that surprised me. 'I've still got sweaty hands though. That remains a fact.'

'Oh, that part is a fact,' she said. 'And it's of no value if you just spout it out there.'

I settled upon the eyes of a middle-aged man called Ron. 'So I'm supposed to be telling Ron how I'm feeling?' I said. 'I can feel my heart beating. My palms are sweating.'

'OK, slow down,' said Paula. 'First of all, you're racing. Just be with Ron in his eyes. Now breathe. I know you're nervous and scared and all these things. Um . . . you're doing something. What's the sensation in your hands?'

I looked down. My arms were rigid by my sides and my fingers were all splayed out, taught and flexing like the legs of a dying spider. It was a completely weird thing to be doing. Oh *God*. 'I think they're trying to crawl off,' I said. There was a scatter of nervous laughter. 'All the way back to England.'

'You've come a long way,' said Paula. 'It's not so easy to pick yourself up. Stay with Ron. What's the sensation in your hands? You said sweating?'

'Yes.'

'OK, move into the sweating.'

'I can feel my pulse in my fingers.'

'Stay with that. Breathe into the sweating. Notice if your mind is chattering away but just keep looking into Ron's eyes and just breathe into the sweating. What other body sensations are you aware of?'

'I feel nauseous in my stomach.'

'But what's the sensation?'

'It reminds me of when I used to drink, when I would be hungover. A kind of poison in my stomach.'

'That's an image. Notice the difference between the label and the actual sensation. There's a sensation you can identify in your tummy.'

I thought about it for a moment. 'Slithering, I suppose.'

Slithering? *Slithering?*

'What?' she said.

'Er, slithering,' I said.

'Slithering?' There was a silence. 'OK. Breathe into the slithering.'

I breathed into the slithering.

'Breathe into the sweaty hands. Don't try to fix them. Just let them hang there. Breathe into the slithering. Good. What else are you aware of?'

'I guess I'm swaying backwards and forwards. Rocking.'

'Stay in Ron's eyes and just notice that rocking. Keep your eyes open. Breathe into the rocking. Notice if the slithering is still there. If so, breathe into it. What else are you aware of?'

'I just feel . . .' My voice sounded thin. 'I just feel completely empty, actually.'

'That's an image. And I'm telling you it manifests as a sensation.'

'OK,' I said, acidly. *I think you're being an arsehole,* I thought to myself. *Please stop being an arsehole.*

'As you feel the emptiness, do you have any feeling about what it's connected to?'

'Er . . . is it connected with something?' Looking out into blinding white light, I felt myself being pulled into the nothing. 'I suppose I might feel a bit like that in my life at the moment,' I said. 'I feel that emptiness of having just turned forty and feeling that my life is quite empty.'

'That's a perception of yourself that you have. You've turned forty and you're asking what have you accomplished? Did anybody ever tell you that you need to judge yourself like that?'

Then, for the first time in many years, and just like that, I was crying. I couldn't believe it was happening. 'Yeah, that is so real, what's coming up for you,' said Paula. 'It's your experience. Breathe.'

I could hardly speak through it. 'I'm just . . . uh . . . I'm just . . . I'm just not sure I'm a very nice person.'

'It's hard,' she said. 'But it's what's lodged in you. Breathe.

Let go of your arms. Keep your eyes open. Look at Ron. Talk about poison. Were you an alcoholic?'

'Yes.'

'Were you a juvenile delinquent?'

'Yes.'

'I'll bet you were a mischievous one. What did you do?'

'Oh, I'd be disruptive in class. I'd want all the attention. And now it's completely changed. I want none of the attention.'

'Because you wanted to disown that guy. You've been busy trying to clean up but you've been repressed and miserable and angry at yourself.'

'I suppose I found out when I was at school. I think I was a bit annoying. I pissed people off.'

'I've got an assignment for you. I want you to be that guy. Fourteen years old. You've pushed him away and he's very much a part of you. That was a strong kid. Rebellious. Did not want to be controlled by fear and terror. But then you decided to clean up and be a good boy. But look, here you are, saying, "My life isn't working. I'm angry. I'm an asshole." So, despite the pasting on of the good behaviour, it doesn't mean anything. Because what you did was disown yourself. Now you get to play at being that kid again.'

I stepped off the stage in a daze. That had gone extremely badly.

That evening, after having successfully dodged the group's attention by arriving twenty minutes late for dinner and finding a spot away from them, I crept into bed with my histories of Esalen and tried to distract myself with tales of Fritz's erection. Hours later, I was still awake and filled with portentous 3 a.m. thoughts. I was dreading my task the next day. I could hardly recognize the loud, attention-loving schoolboy I once was, let alone imagine becoming him. Over the years, I'd somehow managed to become his opposite. These days, I love being by myself. I work by myself, hike by myself, go to the cinema by myself, eat at restaurants by myself, go on holiday by myself.

'By myself' is why I've moved into the countryside, to a quiet cottage at the very end of an old, crumbling private road.

The problem with all this, I've found, is that the more you choose to be alone, the more everyone else wants to leave you alone. Isolation makes you paranoid. Your worst fears about yourself and everyone else fill all that silence you've created, making you more fearful and more crotchety and more averse to human company until you're there, with the blinds down, scowling at the ringing phone, thinking, 'What have I become?' What you've become is a tutter, an eye-roller, a person who sighs loudly in the queue at the shops. Solitude can be an engine that produces its own fuel, sending you faster and faster into the quiet.

And yet, when I was fourteen, I'd crave the company of others. I'd constantly agitate my friends to meet outside Woolworth's on a Saturday afternoon or go drinking stolen amaretto in the woods. When they'd sometimes say no, I'd be mystified. How could you possibly not want to go out? It was fun! It was drama! It was *life*! I had friends, back then, but also plenty of enemies. On at least two occasions, I managed to turn almost everyone I knew against me. I was loud. Disruptive. Annoying.

I remembered one Saturday night that I spent with a crowd of school-friends on a post-cinema trip to McDonald's. There was a scratch-card promotion on, and someone had noticed 'NO PURCHASE NECESSARY' in tiny writing on the bottom of the poster. That meant you could keep asking for scratch cards and win endless fries and milkshakes. It was fantastic, the loophole of dreams. We were sitting around with full bellies and silver fingernails, having taken over several tables, and everyone was looking at me, and I was so happy because I was doing some sort of crazy joke, some sort of *behaviour*, and a girl, who I also happened to be in love with, said, 'Why don't you just go? Nobody wants you here.' I thought, 'She's not serious.' I looked around at the others. I stood up and thought, 'If I walk down the stairs very, very slowly, they'll feel sorry for me. They'll say,

"Come back! We were only joking!"' I walked down the stairs very, very slowly. Then I went home.

When I think of incidents like that (and there are many) an actual noise comes out of my throat. I groan, I flinch, my hands ball, and then I freeze. It can happen anywhere, in the middle of the street, in the supermarket. I just stop, twitch and make a sound, as if the shame is physically coming out of me. It's not only incidents from my youth that sneak up and punch me from the inside, like this. It could be something that happened last week. It's as if the self is turning in on itself, remembering a time when it strayed too far from its model of the ideal, and is punishing itself for the failure.

But as obvious as it was that I'd changed since I'd been that teenager, when I looked back it was still with this weird sense of coherence and continuity. It's not that I doubted Bruce Hood was right when he said the feeling we have a consistent core, like a soul, is an illusion. But I still couldn't shake the idea that there was something in the middle. It was almost as if the different versions of me were different responses to the same problem, as if my mind was trying out different ways of solving it. It had tried loudness and attention-getting, then it tried drinking and drugging and now it was in a phase of dogs and solitude. Perhaps it was the *problem* that was the consistent thing – that was the facet of my self that seemed to have remained stubbornly unchanged. But what was it?

Ah, well, but my left-brain interpreter had a new name for that now, thanks to my conversation with Professor Gordon Flett – neurotic perfectionism. We're those worried and anxious people who have a 'massive discrepancy' between who we are and who we need to be. We make these sweeping generalizations, about ourselves, so if we're not efficient at a particular thing, it's a failure of the entire self. And with this comes a lot of self-loathing.

I feel like I'm breaking some sacred law of our culture by saying what I'm about to – it's an admission that's liable to arouse a kind of contemptuous disgust in many – but here it is. I spend quite a lot of my life in a state of self-loathing. I'm not sure why people don't admit to this more commonly, because I've no doubt it *is* common. Perhaps it's because it (not quite accurately) implies a level of self-pity, and self-pity is a quality that's uniquely unattractive. The feeling of angry repulsion it can inspire when we come across it in others is surely telling. Our storytelling brains want the selves it meets to act as heroes, optimistically rising to the challenges of life. When they encounter the opposite, they react viscerally, as if they've come across something infectious.

But this is a problem. When people feel suicidal it often comes with feelings of self-loathing and, because of this cultural taboo, we don't want to admit them. We certainly don't want to talk about it to anyone, especially if we're prone to perfectionistic thinking and a heightened sensitivity to signals of failure. The way we feel about ourselves is shameful. And in the silence that's pushed upon us, that shame and self-loathing grow.

It seems to me that self-loathing is what happens when our brain's hero-making capacities become defective. When we're happy, we feel good about ourselves, successfully pursuing our meaningful projects, making our lives and the world around us better. We're distracted from the truth of our situation, which is that we have personal flaws that are deep and many, that our lives are ultimately pointless, that we live in a realm of chaos and injustice, and that we and everyone we love are going to die. When our minds fail to distract us sufficiently, all this can seem very close. It can sometimes feel as if we might turn our heads too quickly and actually see the darkness. Even in our most mundane moments – waiting at a traffic light, queuing for ice cream – the hopelessness of it all breathes heavy.

I realize, of course, that 'nihilistic' thoughts such as these are often dismissed as 'adolescent'. But isn't this also telling? It's in our teenage years that the curtains briefly open on these depressing truths, and then we're swept up into the compelling plots of the grown-ups, with all the thrills and responsibilities they bring. Adolescence is the break between the delusions of childhood and the delusions of adulthood, a time when the projects of one phase of life have broken down and the next have yet to emerge. And in that gap we glimpse the horrors that our storytelling brains work so hard to keep from us. I wish my storytelling brain worked better. I *want* some more of that delusion.

Perhaps the thing that's remained consistent is the low self-esteem that underlies neurotic perfectionism. My adult craving for solitude is a response to it now, just as my teenage loudness was then. I wish I knew how to fix it. When I think of high self-esteem I imagine a golden city on top of a hill. I've been trying to get there for years and, if I was honest, I doubted tomorrow's shenanigans would be of much help. It was going to be a challenge. 'What you did was disown yourself,' Paula told me earlier. She was right about that. Finding the noisy, company-loving young man I once was wouldn't only mean opening myself up to fresh humiliations, it would mean doing it amongst *people*.

The next morning, I began. In accordance with Fritz Perl's gestalt method, Paula had instructed many of us to inhabit, fully and without reserve, the version of ourselves that we feared the most. One woman had to be an authoritarian police officer, and prowled the Institute in mirrored sunglasses, snarling and barking orders. Another had to be invisible and floated about mutely in dark glasses with bandages over her head. A man in his thirties had to be his own volatile Vietnam veteran father, who'd drawn a gun on him when he was three. He spent his time shouting and interrupting everyone. ('Owls? I fucking hate 'em.') In the dressing-up box, in the laundry area, I found an old ripped Metallica T-shirt. Watching me, from a distance, was the police officer.

'You're gonna put that on right now, motherfucker,' she snarled. She had an eighteen-inch truncheon and was slapping it against her open palm. I did as I was told. Outside I saw a large laundry cart. It gave me an idea.

'Officer,' I said to her. 'Could I offer you a lift down to the Big Yurt?'

She laughed, delighted. 'You bet!'

She jumped into the cart and I began rolling her down the hill. As gravity took its effect and we ran faster and faster, bumping and giggling over the sun-speckled tarmac, I felt suddenly joyful and stupid and full of life. It was astonishing how utterly I became that fourteen-year-old again. When we got there, I parked up and let her climb out. Out of the corner of my eye, I noticed what seemed to be a goat. 'Brilliant!' I said, now delirious with it all. Forty seconds later, cheered on by the others, I had untied the goat and was leading it away towards a fruit patch, where I planned to share with it a delicious feast of stolen organic strawberries.

'Excuse me!' came a voice. 'Hey. *Hey!* You're going to need to bring that goat back right now.'

I turned. There was look in the goat woman's eyes that I also recognized from my youth. 'It's for the kids,' she said, with exasperated disappointment.

'I'm sorry,' I said. 'I'm so sorry.'

I walked the goat slowly back, feeling ridiculous in my ripped T-shirt. I knew exactly what that fourteen-year-old would've done in a circumstance like this. He'd have said, 'Fuck this.'

The next day we were instructed to update Paula on our progress. When it was my turn, I shuffled onto the stage, blinking under the light.

'A couple of people thought you've backed off your assignment,' she said.

'I stole a goat!' I protested weakly.

There was a voice from the audience: 'Will told me the cheekiest thing he could do would be to not do his assignment.'

This was true. Paula studied me sadly. 'It would be hell to

think you came all this way just to go back to your cave,' she said. 'What are you afraid of? Getting to know people? What do you fear?'

'As I'm getting older, I'm getting more grumpy. I don't know . . . I'm just not very good at getting on with people. I don't know why.'

'Well, how about that for an assignment? The guy who doesn't get on with people.'

'To be a prick?' I asked.

'To be a prick.'

On the way out of the Big Yurt, I passed a heavyset woman in her late fifties whose task was to be a cavewoman. She was in a tree, crouched on a low branch, grunting and honking, her left breast hanging out of the crude clothing she'd made from tied rags. It was the same woman who'd snitched on me to Paula for not doing my task. She leapt to the ground, squatted in a flowerbed and began to urinate, the dark pool curling and sinking into the soil. This, I realized, was my moment.

'What the fuck do you think you're doing?'

'Grfghhsgh hugghh,' she said.

'You're disgusting,' I told her.

'Hugrrrgghh hghghg.'

'You're embarrassing yourself. Do you understand that?'

'Grrrhhrggh. Hugghhhsss.'

I leaned towards her, right up to her face.

'Put some clothes on. Shave your armpits. And grow the fuck up.'

'Grfghhsgh. Hghghg.'

This was the me I feared the most. He was the lonely man, the angry man, the weirdo. He was the cunt. And, in that moment, I had a terrible realization. I was *loving* being the cunt. I was loving being the cunt so much, I didn't think I ever wanted to stop.

*

It was the early 1970s and Esalen was successfully pumping its brain into America's middle class. There were nearly one hundred unaffiliated so-called 'Little Esalens' around the country and the Institute's work was being seriously discussed in some of the most prestigious halls of learning: Stanford, Harvard, Berkeley, UCLA. Thousands of the nation's psychiatrists, social workers and clinical psychologists were coming to find their authentic selves before travelling back across the States to practise what they'd learned. Esalen had opened an outpost in San Francisco which reportedly received ten thousand people in its first two months. Schutz was a celebrity. The hit film *Bob & Carol & Ted & Alice*, starring the unhappily spanked Natalie Wood, satirized the Californian hot-tubs of personal transformation ('The truth is always beautiful!'). It was now pouring into the nation at large: the privileged middle classes' discovery of recreational yoga, massage and meditation as ways of purifying and bringing peace to their once-Christian interiors; chatter about authenticity and auras; earnest deconstructions of personal relationships; little Esalen phrases – 'I love your energy'; 'Go into the pain'; 'I hear you'; 'Be real'. It had its cauldron on this Big Sur cliff.

But at the Institute itself, things were changing. In the media, founders Murphy and Price began downplaying the promises of catharsis and radical transformation that had become its trademark. The programme of lengthy residential courses, which had seen some of Esalen's most berserk scenes, was halted. 'Many reasons were given for ending it,' writes Anderson, 'but the overwhelming reason seemed to be that altogether too many of the people in it ended up committing suicide.' The deaths involved men and women who had been in Esalen's constellation, near and far. Whilst it would be wrong to blame the Institute directly for what was happening, it was nonetheless troubling. In 1968, Marcia Price, who'd been a patient and a lover of Fritz, was found in a Volkswagen camper, which was parked on the grounds, having shot herself through the head with a rifle. After her death, footage was released of a Gestalt session in which Fritz had taunted her with her own suicide threats. Another vulnerable woman mocked by Fritz was a psychologist named Judith Gold. She'd confessed to having had suicidal thoughts and was 'degraded aloud and mocked by Fritz', according to witness Jacqueline Doyle. She left the group 'distraught'.

The next morning Judith Gold drowned herself in the baths. Back in the group, 'everyone was totally frightened and spooked by this, and there was a lot of confusion of feelings toward Fritz,' reported the witness. 'Fritz was being very offhand and callous, no grief being expressed. Just "Ach, people who play games." You know, his way.' In 1970 an attendee named Sunshine shot himself in a barn. Then a Harvard graduate named Nick Gagarin, who'd written about Esalen several times in the *Harvard Crimson* before attending a four-month residency, shot himself at his father's house. Another former member of that programme, Jeannie Butler, apparently threw herself into the Pacific: her clothes were found on the edge of the cliff. Art Rogers, the shy psychotherapist who was chased and attacked in Schutz's group, would also go on to kill himself. And if all that wasn't enough, like some strumming, soothsaying demon, Charles Manson dropped by to play some tunes just three days before the mass-murder that would make him famous, and would mark, for many, the spiritual end of the 1960s.

*

The lack of a perfect, authentic self to actually uncover during Esalen's encounter groups wasn't the only flaw in their model. Far more dangerous than that was the fact that what Rogers, Perls and Schutz had unwittingly created was a perfect environment in which to generate a species of agony whose true nature was not well understood at the time. It's only in recent years that scientists have begun to properly understand what's known as 'social pain', which is that which springs from rejection and ostracism. Their work illuminates fresh parts of the human self that are common to each one of us.

The psychologist Professor Kip Williams had a memorable, and as it turned out, consequential, experience of social pain one afternoon during a picnic. He was kicking back with his dog Michelob, near a public lake in Des Moines, Iowa, when a Frisbee landed beside them. Kip looked up to see two men waiting for it to be thrown back. 'So I got up and threw it to them,' he told me. 'And I was going to sit down again but, to my surprise, they threw it back to me. We just started throwing the Frisbee to each other.' The game continued happily for a while. But then something banal and terrible took place. 'They just stopped throwing to me,' he said. 'They didn't say anything. They just started looking at each other and not at me.' Kip was left just standing there, in front of Michelob, feeling awful. 'I was amazed at how powerful this fairly minimal experience of ostracism was. I felt it in a visceral way. I felt it in my gut. I hurt.'

What the men didn't know was that Kip was a social scientist who'd been looking for a way to study ostracism. They'd given him a brilliant idea. He decided to recreate the situation in the lab. He'd record what happened when a person was engaged by two strangers in a catch game before suddenly being frozen out. 'It had an extraordinary impact on them,' he told me. 'It affected their self-esteem, their sense of control over their environment and what we call their "meaningful existence", which is whether someone feels acknowledged or invisible. It also affected their anger and their sadness.' Kip observed some of these experiments through a one-way mirror. 'They were so powerful I had trouble watching.'

All this undermines the folk truth that 'names can never hurt me'. They can and they do. 'People say social pain "is all in your head" and, indeed, it is,' said Kip, 'because that's where you register both physical and social pain.' (Some researchers, in fact, believe they actually use the same brain networks but, at the time of writing, a dispute has broken out in academia as to the extent of the overlap.) And social pain can be no less painful. 'Sometimes we'd prefer to experience physical pain than the breakdown of a relationship,' Dr Giorgia Silani of Italy's International School for Advanced Studies told me. 'You feel that kind of pain in your body. It's like your body is getting sick. This is what we're picking up.' Her team ran a series of studies using ostracized catch players in a brain scanner, except this time they were also given electric shocks, so that a comparison could be made between them. 'We found social pain could be as strongly felt as physical pain.'

There are many different kinds of social pain: embarrassment, betrayal, bereavement, insult, exclusion by a group or individual, loneliness, heartbreak. What they have in common is rejection. Ostracism is a capital assault on the self, sometimes described as 'psychological death'. (It's not for no reason that St Benedict considered 'the cauterizing iron of excommunication' the ultimate punishment for wayward monks.) The reason we've evolved to experience it with such agony is thought to go back to when we roamed the planet in vulnerable tribes. 'The tribe is providing you with protection and with food and water,' said Giorgia. 'To hunt, you need maybe five or six people. You can't do it by yourself.' If you were rejected by the group, back then, it would very probably mean death.

This is perhaps why social pain developed: it was an alarm system that told you something was wrong in your social life and that you needed to take urgent action. In this respect, it's no different from physical pain, which works as an alarm system that tells you not to touch an open wound, say, or walk on a broken ankle. Pain is information. 'There are people born without the ability to experience physical pain,' said Kip. 'They typically die by the time they're in their mid-twenties.' Today, some researchers believe social isolation to be so bad for the human machine that it can rival smoking as a mortality risk.

But social pain is more than just an alarm for ourselves. We also experience it when we see *someone else* being ostracized. Giorgia worked on a series of studies that examined people's responses to watching another be excluded. 'What we saw was a reactivation of the area that we observed in the first-person experience of pain,' she says. 'So it was just as painful to watch the experience happening to someone else as it was to experience it yourself.' This we sometimes call 'empathy'. It's argued that we feel pain on behalf of others because in order to keep tribal life ticking over, we'd have to have been motivated to punish those who were treating others badly. 'In a social environment, if you observe unfairness, you'd probably try to punish that kind of behaviour so it doesn't happen again,' she said.

But we don't feel empathy for just anyone. Professor James Coan at the University of Virginia has suggested that we only tend to experience it on behalf of people who are members of our in-group. His team reported that when participants in an fMRI scanner believed a friend was about to receive an electric shock, brain regions that are typically involved in threat response became more active, just as they did in Giorgia's tests. But when that threat was made to a stranger, only minimal activity was observed.

Other studies, by researchers at China's Shenzhen University, have indicated that we can struggle to feel empathy for those who we think of as having a higher status than us. This effect, of course, is made manifest in the fact that we often feel entitled to be extremely bullying and unfair to politicians, CEOs and celebrities who are, after all, no less human than we are.

Once moral outrage is triggered, so is our thirst for revenge. Our use of ostracism as a weapon of assault is with us as much today as it's always been, of course, and the science of social pain suggests it can sometimes feel hardly less brutal than the physical variety. 'Anthropologists think ostracism was the foundation of civilization, because fear of it keeps you in line,' said

Kip. 'But taken too far, it makes everybody too similar to each other. It penalizes diversity and creativity. You end up wanting to get along to such an extreme that you fear expressing anything unique.' In today's culture, this effect can be seen often in social media, newspapers and on university campuses. 'You see it on both sides, from the right and left. There are strong pressures to conform and an immediate response to disrupt or to ostracise people who disagree.'

What the science of social pain begins to reveal is the dark error at the heart of individualism. We're not the solitary apes of cliché but a species so highly social that psychologist Professor Jonathan Haidt calls us '10 per cent bee'. We evolved to thrive in communities, not out there on the plains alone, like that cultural icon of individualism the American cowboy. But when our cultural ancestors in Greece decided that the world, including people, was made up of individual objects, they unwittingly averted their gaze from our natural state of interconnectedness. The American self brought with it a wildly magnified emphasis on the power of the individual. The self was inherently heroic, now, and if you failed to live up to the heroism that was already in you, you were categorically a failure. The age of perfectionism was coming.

*

The next day, walking to the Big Yurt, I felt as if my entire upper body was built from electricity. Through the dangerous crackle, I could see four of the worst hippies waiting on the seats by the entrance.

'Hey, Wiiiilllll!' said one of them, waving her open palm in a semicircle.

'Fuck yourself,' I said, sitting down next to her.

'It's a beautiful morning.'

'Another day in twat paradise.'

She grinned at me. 'You're not fooling us, my friend. You're not fooling anyone.'

'And I'm not trying to.'

'So, Will, you gotta tell us. We've all been talking about it. Why the hell do you think people don't like you?' said the standing woman to my left. 'You're sweet. You're funny. You can't hide it. Everybody here likes you.'

'Well, I hate you.'

She started laughing. 'You're being funny *now*.'

'No, I actually do.'

She looked me in the eyes and smiled, then reached out and rubbed my shoulder. I couldn't work out if they were being honest or not. They seemed nice, anyway. They seemed like good people. I stared at the floor. My throat felt swollen.

Inside the Yurt, Paula issued our final task. In small groups, we were to write and rehearse a short skit that we'd perform on the stage. This meant it was the end of my 'being a prick' assignment. And, for that, I was glad. At first it had felt joyous to finally be able to say what I was thinking without fear of judgement or censure. Living the arsehole me, in full gestalt mode, had been fun. But it had also been surprisingly therapeutic. Having permission to be that person I didn't like so completely had somehow taken his power away. That version of me I'd been so frightened of came to seem like just a lot of wind and insults that quickly blows itself out. I'd given him the chance to take me over, and he'd failed. And even better, my new understanding of the science of self told me that he can't be the 'real me', because there *is no such thing*. The element that felt like such a constant within me, I decided, was surely my low self-esteem. I just needed to try a bit harder, that was all. And I could start by opening up to these people.

That evening, I sat down to dinner with the women who were in my performance group. We discussed the slight let-down we'd felt at discovering the notorious Max would peak with nothing more than a theatrical turn on the stage. 'I know she's been doing this for *over thirty years*,' said one of them. 'But I thought we were going to go further and come out the other side, and that's when it would happen.' She took a bite of her gluten-free spinach pie. 'Maybe the transformation comes when we get the fuck away from her.' I collapsed into laughter, and so did everyone else. I realized I was completely surrounded by people from my course. I felt immensely happy to be with them. I didn't want to leave. For the next day and a half, as our group wrote, rehearsed and performed our appalling skit, I can honestly say that I haven't laughed so much or so freely since I was a boy. The sun shone. The waves crashed. It was magical.

On the morning of my departure, I made my bed and vacated my room at the appointed time, so that it could be cleaned for the next guest. I had a heavy bag, and four hours to kill, so I carried it to the reception desk to be locked away.

'Could I please leave my bag back here until the bus comes?' I asked the doe-eyed young man at the counter. He gazed at me beatifically. 'We don't take that responsibility.'

'Right,' I said. 'Sure.'

I dragged myself and my bag out onto that perfect lawn, overlooking the cliffs and the ocean. When it was finally time to leave, I became aware of a gaggle of youngsters up by the pool, singing 'Swing Low, Sweet Chariot'. The men had long hair and their tops off and the braless women threw shapes with their arms and raised their faces to the sun. There were bongos. When I heard them briefly talking I learned that one of them was called Flowers. To my pleasure and relief, I found myself free of the scratchy middle-aged voice in my head that'd usually be there, judging and cursing these joyous young people. I couldn't help but smile. Maybe Esalen had worked. Maybe I'd actually changed. Then, as I pushed myself to my feet, without any warning at all, I heard myself mutter, 'Fucking idiots.'

*

In the months leading up to his death, in 1970, the humanistic psychologist Abraham Maslow began worrying about his legacy. He'd been preparing to write a critique of Esalen 'and its whole chain'. One of the issues he'd become concerned with was self-esteem. Maslow was famous, most of all, for his hugely influential 'Hierarchy of Needs', which said that people are motivated to fulfil certain psychological appetites. At the top of his pyramid was 'actualization', which was extremely difficult and had, he thought, only been achieved by a few. But just beneath that was 'esteem'. It seems that Maslow had been carrying out some tests on high-esteeming people that had been the cause of some concern: 'High scorers in my test of dominance feeling, or self-esteem, were more apt to come late to appointments with the experimenter, to be less respectful, more casual, more forward, more condescending, less tense, anxious and worried, more apt to accept an offered cigarette, much more apt to make themselves comfortable without bidding or invitation.'

He wasn't the only Human Potential guru who'd apparently come to a place of doubt late in life. According to Dr William Coulson, the chief of staff at Rogers' Western Behavioral Sciences Institute, 'After several years in California, Carl got so tired of having aspirants arrive at our door with no intention or ability to WORK that he sent out a letter. It said, in fact, "Less self-esteem, please. More self-discipline."'

In December 1973, Esalen hosted a conference in San Francisco called 'Spiritual and Therapeutic Tyranny: The Willingness to Submit'. The Institute's co-founder, Michael Murphy, had grown troubled, both by his observations of a cultish, guru-worshipping aspect that was developing in some quarters and also by new commercialized forms of personal transformation workshops, such as EST, which took place over two weekends and combined Esalen-style Human Potential thinking with a sales seminar format more familiar in an earlier period of America's history. EST was created in 1971 by Werner Erhard, an Esalen graduate and student of both Carl Rogers and Abraham Maslow, who had also been a fan of a Dale Carnegie course he'd attended. Erhard, writes Truett, 'Americanized the human potential movement in a way that Esalen had never done or tried to do.' It was an instant success: fifty thousand people took the workshop in its first four years and it ended up boasting celebrity attendees including John Denver, Cher and Peter Gabriel. It represented a kind of business-ification of Human Potential methods that Murphy hoped to distance Esalen from with his conference.

It didn't go well. Despite being organized by a mostly female committee, all twenty-six speakers were male – which led to protests, outside the venue, by feminists from Esalen's San Francisco office. A journalist covering the event, Peter Marin of *Harper's Magazine*, described the crowd of several hundred attendees as 'restless, impatient, volatile; one could feel rising from it a palpable sense of hunger, as if these people had somehow been failed by both the world and their therapies.' They had come to the conference, he wrote, for the same reason they had attended the workshops: 'to find help. The human potential movement had still not done for them what it had promised, their lives had remained the same or perhaps worsened, and the new world, the promised transformation, seemed very slow in coming.' As the event ground on, the audience barracked the panellists and the panellists shouted at each other. In his keynote address, the writer Sam Keen pointed out, devastatingly, that the 'usually male, privileged' leaders of the movement had hardly demonstrated, in their own selves, the efficacy of what they preached. 'The best therapist turns out to have a clay heart. And Fritz was a dirty old man. And Freud couldn't give up cigars. And Bill Schutz doesn't jump for joy.' It was a criticism redolent of an observation made by a former lover of Michael Murphy, who had himself suffered lifelong mental-health problems: 'He got a million dollars' worth of advice from some of the best psychologists in the country. None of it helped.'

Peter Marin's report on the conference would end up being part of a cover story for *Harper's*, published in October 1975. He captured what happened when Christian introspection met Human Potential's belief that perfection, not evil, lay in our souls. The speakers, wrote Marin, exhibited 'a tyrannical refusal to acknowledge the existence of a world larger than the self' and when the audience questioned the panellists, 'the questions they asked were invariably concerned with themselves, were about self-denial or self-esteem, all centred on the ego, all turned inwards'. The Human Potential Movement had posed the Western self a question: if God is inside all of us, then doesn't it naturally follow that we are all Gods? Now the Western self had given its answer. Marin called his story 'The New Narcissism'. It described the stunning darkness of Esalen's conviction that, as gods, all humans have complete responsibility for everything that happens to them, even including the Jews that 'burned' in the Holocaust. When Marin asked one counsellor, at the institute, whether we owe anything to a child starving in the African desert, she snapped angrily back at him, 'What can I do if a child is determined to starve?'

Ten months later, America's chief chronicler of culture, Tom Wolfe, would author his own cover story on the subject, this time in *New York Magazine*. He named Esalen 'Lemon Session Central' and described the programmes of Fritz and Schutz in wild and cynical terms. 'Outsiders, hearing of these sessions, wondered what on earth their appeal was,' he wrote. 'Yet the appeal was simple enough. It is summed up in the notion: "Let's talk about Me." No matter whether you managed to renovate your personality through encounter sessions or not, you had finally focused your attention and your energies on the most fascinating subject on earth: Me.' Wolfe's piece was called 'The "ME" Decade'.

The old-age concerns of Maslow and Rogers had come too late. Their bright ideas had run out of control. But this feel-good American revolution, in which the interior self was recast as holy, and took responsibility for everything that happened to it, would turn out to be perfect for the coming times. In the 1970s and 1980s, as the economy underwent new, dramatic changes, the English-speaking nations became feverish with the idea that all of society's problems, from unemployment to child abuse to domestic violence, could be solved by teaching the people to believe in their authentic, godlike selves. This harsh new world would be no place for 'fakes' or 'shitheads' or 'weepers'. Everyone was special, and had what it took to thrive. All they had to do was *believe.*

One of the invited speakers, at Esalen's terrible conference, was an imposing and angry politician with a deep, wounded-bear voice, who'd had his life changed partly by his experiences at the Institute, having first visited in 1962. He would help take Esalen's stated mission – 'that all men somehow possess a divine potentiality; that ways may be worked out – specific, systematic ways – to help, not the few, but the many toward a vastly expanded capacity to learn, to love, to feel deeply, to create' – into the 1980s and beyond . . . into schools and prisons and courts; into government policy and around the world. He too would be dogged by the imperishable charge that all he was doing was indulging and encouraging narcissism.

He was a reformed Catholic, a disciple of Carl Rogers, an Esalen alumnus and soon to become one of the most powerful men in California. He was John 'Vasco' Vasconcellos and the mission of his life would be to give the world self-esteem.

BOOK FIVE

The Special Self

So there was this girl, name of Alyssa Rosenbaum, who spent her childhood in St Petersburg. She was born in 1905 into a bourgeois family and then, when she was twelve, the Bolsheviks came. They confiscated her father's chemist shop and forced the family to flee the city, poor and starving. Even at that young age, she detested the extreme collectivism that had become rampant in her country. 'I thought, right then, that this idea was evil,' she said. 'I was already an individualist.' Alyssa came to America, where she found celebrity and notoriety, and would eventually change the nation's sense of self. Her influence is still with us today, with separate veins running through the school system, the global economy, Silicon Valley and the halls of government. In many respects, in the second decade of the twenty-first century, we're all now living in Alyssa Rosenbaum's world.

After a spell in Chicago, Alyssa moved to California, where she found work in Hollywood as an extra, a wardrobe assistant and then a screenwriter. She began writing books that raged against the collectivist spirit of the Great Compression that had overcome the US. In 1943, with the publication of her third novel, she found worldwide fame. The book was a hymn to individualism, a moral argument that human civilization was the work of single-minded 'creators' who needed, more than anything, to be free in order to build. The opposite of creation, she thought, was altruism. Alyssa hated altruism. 'Men have been taught that the highest virtue is not to achieve, but to give. Yet one cannot give that which has not been created.' On the contrary, she believed that people should always put themselves first. 'The first right on earth is the right of the ego.' By this time, Alyssa had changed her identity. She'd combined the name of a Finnish writer with the make of her typewriter and was now calling herself Ayn Rand. Her bestselling book was *The Fountainhead*.

In 1951 Rand and her husband moved to New York. As she laboured on the follow-up to *The Fountainhead*, a small group of acolytes gathered around her. In a knowing smirk at the commitment to individualism that ironically bound them together, she named the group 'The Collective'. But her joke had an even heavier layer of irony bolted onto it. In reality, the Collective was a kind of cult that operated under a system of sacred beliefs which, one of its members would eventually admit, included 'Ayn Rand is the greatest human being who ever lived' and 'Ayn Rand, by virtue of her philosophical genius, is the supreme arbiter in any issue pertaining to what is rational, moral, or appropriate to man's life on earth'. They began to see themselves as pioneers whose pursuit of Rand's vision of 'virtuous selfishness' would dismantle the calamitous Great Compression, and replace it with a world of small government, minimal taxation, deregulation and free markets in which 'man' would be free to compete with 'man'. 'We thought of ourselves as being the instigators of a revolution that was coming,' one member recalled, in a 2010 interview with the filmmaker Adam Curtis. 'We were so enthusiastic about what a difference this was going to make to the world.' When Curtis asked what they hoped to achieve, she replied, 'A totally free society.'

Rand claimed that the only thinker to have ever influenced her was the father of individualism, Aristotle. It was he who, more than two millennia earlier, had decided that a state of 'ennobled self-love' was a precondition for the successful pursuit of perfection. In order to achieve, people first had to love themselves. 'Man is entitled to his own happiness and he must achieve it himself,' she said during an appearance on ABC's *The Mike Wallace Interview*, her accent thickly Russian, her words thickly American. 'Nor should he wish to sacrifice himself for the happiness of others. I hold that man should have self-esteem.'

Rand's belief in self-esteem spread widely around the US, Britain and elsewhere. It's still with us today. But the idea's popularization began via the work of the man she called her 'intellectual heir', a senior figure in the Collective who, despite being married to another member, and being twenty-five years Rand's junior, was also her lover. Nathaniel Branden co-edited and wrote essays for Rand's 'journal of ideas', *The Objectivist*. His 1969 book, *The Psychology of Self-Esteem*, was the first of its kind, selling more than a million copies, and informing the schools programmes that were established in the years following its publication. Based principally on his writings in *The Objectivist*, he argued that self-esteem, founded in rationality and achievement, was the most important factor in a person's psychological development. 'The nature of his self-evaluation has profound effects on a man's thinking processes, emotions, desires, values and goals,' he wrote. 'It is the single most significant key to his behaviour.' During the 1970s, Branden became known as 'the father of the self-esteem movement'. But his influence extended a great deal further when he both inspired and worked directly with the people around California Assemblyman John Vasconcellos, who in the 1980s and 1990s did more than anyone else to take the ideology of self-esteem into lives all over the world.

If Rand was influencing millions directly with her best-selling novels, and Branden was taking her ideas of self-interest and self-esteem into schools and therapy rooms, there was a third member of the Collective who grew to have an even greater influence yet. She met him in New York when, every Saturday night, the group gathered at her apartment to hear the latest excerpts from her novel-in-progress, *Atlas Shrugged*. This was to be her masterpiece, her attempt, in the words of biographer Anne Heller, 'to create an ideal man and delineate the idea and worldly conditions that would allow him to live, love, create and produce.' *Atlas Shrugged* imagined a dystopian America entirely controlled by the state, in which the creators – artists, industrialists, entrepreneurs – rebelled to form a new world for themselves in a secret, far-away valley called 'Galt's Gulch'. It was a place of reason, the quality Rand admired above all others. Only once rationality had freed you from the shackles of emotion could you be a productive member of this utopia in which no one paid tax, everybody competed with everyone else and the markets were unhindered by regulation – after all, just as people needed to be free, so did the markets they naturally created. As Rand once said, 'a free mind and a free market are corollaries'.

One member of the Collective who became notably exhilarated by these Saturday-night readings was an earnest twenty-six-year-old named Alan Greenspan. With his glum-doggy countenance, quiet manner and thundercloud suits, Rand gave him the nickname the Undertaker. It was only when he first read excerpts of *Atlas Shrugged* that another side of him emerged. 'He came alive with an excitement that no one had ever seen in him before,' remembered Branden. Greenspan thought her ideas 'radiantly exact' and so perfect in their logic that those who claimed to disagree must surely be lying. 'What she did,' he said, 'was to make me see that capitalism is not only efficient and practical, but also moral.' She also taught him about the fundamental nature of humans, 'their values, how they work, what they do and why they do it, and how they think and why they think'. From being something of an outsider, he began helping Rand with her book, advising her on the economics of the steel industry. Having now decided he had a 'first-rate mind' Rand became so convinced he would live a life of significance she changed his nickname from the Undertaker to the Sleeping Giant. He contributed his own essays to *The Objectivist*, in which he argued that the markets were benevolent and self-correcting: 'It is precisely the "greed" of the businessman, or, more appropriately, his profit-seeking, which is the excelled protector of the consumer.'

Atlas Shrugged was published in October 1957, when the cultural power of the Great Compression was mighty. The fashionable critics were savage. Gore Vidal called it 'nearly perfect in its immorality', the *New York Times* said it was 'a demonstrative act rather than a literary model' and 'written out of hate' whilst the *National Review* said, 'From almost any page of *Atlas Shrugged*, a voice can be heard, from painful necessity, commanding, "To a gas chamber – go!"' (The implication of this, at least, was unfair. For Rand, appeals to one's identity group were anti-individualistic. Racism, she believed, was 'the lowest, most crudely primitive form of collectivism'.) In an uncharacteristic fit of altruism, the Collective rallied around their devastated idol. Branden instructed them to begin a letter campaign to the worst of her critics. Alan Greenspan wrote to the *New York Times*, arguing that *Atlas Shrugged* was not hateful, as they'd written, but rather 'a celebration of life and happiness. Justice is unrelenting. Creative individuals and undeviating purpose and rationality achieve joy and fulfilment. Parasites who persistently avoid either purpose or reason perish as they should.'

But the reaction of the public was different. It's testament, surely, to the deep individualist core in the American self that even in these collective times, *Atlas Shrugged* entered the *New York Times* bestseller list three days after its publication and stayed there for twenty-two weeks. But it was of little succour to Rand. The critical response ripped her into a state of enduring depression. It affected her relationship with Branden, with whom, over the next two years, she'd have sex fewer than a dozen times. On the evening of 23 August 1968, Rand discovered he'd been seeing someone else. She reacted with vengeful rage, hitting him three times across the face and screaming, 'You have rejected me? You have *dared* to reject me?' She publicly named him a 'traitor' and falsely accused him of a series of misdeeds, including financial malpractice and 'sordid' behaviours. Her denunciation, published in *The Objectivist*, ran to fifty-three paragraphs over six pages. His banishment was approved and signed-off by her loyal Collectivists, including Alan Greenspan. At one point, word reached Branden that the Collective had engaged in one of their theoretical, rational discussions: 'Would it be ethical to assassinate him in light of the great pain he had caused Ayn Rand?' They apparently decided it would.

But Rand and her Collective were to have their time. The years of 'class compromise' between capital and labour that characterized the Great Compression had been sustained, in part, by a kind of virtuous circle. The US economy was one of mass production, and, partly because of the protections afforded them by unions and the state, the wages of the middle classes had been high enough to buy what the nation itself had been making. But then in the 1970s, in the US and the UK, everything started to go wrong. The economy stagnated, inflation surged, stock markets crashed, there was the oil crisis, steel crisis, banking crisis, the 'Nixon Shock', the three-day week. The GDP tanked, the unions fought, millions lost their jobs. It was during these tumultuous times that Greenspan began manoeuvring himself into a position of enormous power. He'd entered politics in 1968, after insistent urging by Rand, as an adviser to Richard Nixon. By 1974 he'd been made Chairman of the Council of Economic Advisers, with Rand proudly watching over his inauguration. From these powerful offices, he'd watched the collapse of the old world and the birth of the new. As the sureties that girded the Great Compression began falling from the sky, politicians desperately sought a fresh theory by which they could run their economies and nations. The solution they placed their faith in eventually led to the long-unfashionable ideas of Rand and Greenspan becoming ascendant.

The idea that quickly emerged as a favourite, and still dominates much of the world today, was 'neoliberalism'. This once-maligned concept is most commonly associated with the Austrian economist Friedrich Hayek. In earlier decades, traumatized by the rise of fascism in his formerly civilized nation, Hayek noticed that what both the Nazis and the Communists had in common was their attempts to control the world via central planning. When he saw the same thing happening in the US and UK during the era of the Great Compression, it horrified him. 'There is more than a superficial similarity between the trend of thought in Germany during and after the last war and the present current ideas in the democracies,' he wrote, in 1944.

From his position as a lecturer at the London School of Economics (where, with his thick accent, he earned the mocking nickname 'Mr Fluctooations'), and subsequently at the University of Chicago in the 1950s, Hayek excoriated the Great Compression projects of Britain and America as being a betrayal of their Ancient Greek inheritance. Central planning was incompatible with the individual freedom on which these great nations had been founded. It had put us on a 'road to serfdom'. What troubled him, in particular, was state interference in the activity of the markets. Hayek argued that those who put themselves in charge of the money put themselves in charge of everything. 'Economic control is not merely control of a sector of human life which can be separated from the rest,' he wrote. 'It is the control of the means for all our ends.'

His dream was to create a world in which the 'coercion of some by others is reduced as much as possible'. In order to achieve this, and avert the slide into totalitarianism, this meant reducing the reach of governments. If individuals were to remain free, and the horrors of communism and fascism avoided, the power of the state would have to be reined in. Rather than flawed ideologues being in charge of nations, control would be transferred to the markets, which would be freed as much as possible from state interference. These free markets would be the engines of a new kind of society in which everything possible would be reconfigured around the principle of competition. The human world would become a kind of game in which we'd all compete with each other, with the deserving winners taking the spoils. These super-rich victors would be heroic pioneers. Having been empowered to create massive wealth, they'd 'perform a necessary service' by 'experimenting with new styles of living not yet accessible to the poor' and thereby forge our futures. All this would put individualism back, where it belonged, at the heart of Western society. For Hayek, neoliberalism was the ideology of no ideology: it would bring a utopia in which we'd finally be liberated from the follies of the politicians.

It wasn't until the 1970s, when the old system began to collapse, that neoliberalism was sucked dramatically into the mainstream. With the help of a band of powerful businessmen, thinkers and economists known as the Mont Pelerin Society, it had grown, since the 1940s, spreading through a network of well-funded 'think tanks', to eventually become influential in all the right places. In 1975, Hayek met Margaret Thatcher at a meeting organized by the Institute of Economic Affairs and was apparently starstruck. His first comment on her departure was, 'she's so beautiful'. As for Thatcher, in her memoirs she calls Hayek's book *The Road to Serfdom* 'the most powerful critique of socialist planning and the socialist state' and admits returning to it 'often'. Soon, neoliberalism was taken up as the guiding principle both of her government and that of Ronald Reagan. It also, of course, had much in common with the views of Ayn Rand and her acolyte Alan Greenspan – who, in this new world, suddenly found himself in favour, and in power.

And so, with Thatcher famously declaring, 'There is no such thing as society. There are only individual men and women and families,' the two world leaders took up their mission to liberate the individual from the shackles of the oversized state and turn society into a game of warring individuals. They would increase competition as much as and wherever possible.

Everyone would compete in self-correcting, wealth-creating free markets (after all, just as humans required freedom to fully actualize, so too did markets) that the state would manage and whose 'invisible hand' could be relied upon to raise us all into a stable and wealthy future.

In June 1987, Ronald Reagan gladly announced the appointment of his new Chairman of the Federal Reserve – Alan Greenspan. This position, writes economist Dr E. Ray Bradbury, made Greenspan 'the single most powerful figure affecting the global economy'. He remained in this extraordinarily powerful job until 2006, a period of nearly thirty years, during which he became known, by some, as the 'Central Banker of Neoliberalism'.

The neoliberal revolution brought with it a new definition of government. The state would no longer be a mechanism for paternalistic control. Instead it would act as both groundskeeper and referee for the great game, being responsible for maximizing and policing the freedom it demanded. In order to encourage competition, it would deregulate business and banking; sell off and release into the markets its utilities – telecommunications, water, electricity, gas – and, in the UK, council houses; cut arts funding; dismantle protections for those not willing to work, thereby gifting them the motivation to get up and join in; cut taxes for the creators – the entrepreneurs, industrialists and their corporations – which would reward them for their superior gameplay and enable them to compete all the better.

For this new style of government, there would be no more 'citizens', but 'clients' or 'customers', whose votes you'd collect like coins in the till. The era of mass production would end. The unions would be fought and conquered. And up would rise, from the dust of mines and ruins of factories, a new army of skilled, flexible entrepreneurial individuals who were free to make their own choices, and captain their own lives, all in healthy competition with each other.

And the revolution would be global. The state's legal and

military powers would be used to protect the smooth running of both local and international markets. The dream would be of a single competitive world market with no barriers to trade or controls on financial flows. The World Bank and the IMF would assist in this mission by offering desperately needed loans to developing nations on the condition they introduced neoliberal reforms. An era of 'globalization' would begin, of money and services zipping around the world, of cheap migrant labour, of manufacturing and service industries being relocated to poorer, cheaper countries, and of financial institutions and multinational corporations that would grow so vast and powerful they'd eventually rival nation states.

But the ultimate goal of this neoliberal project was about much more than ending the economic chaos of the 1970s. It was the creation of a new form of human. Discussing her plans in a 1981 interview, she said, 'What's irritated me about the whole direction of politics in the last thirty years is that it's always been towards the collectivist society.' Then she made a comment that bordered on the sinister. 'Economics are the method,' she said, 'but the object is to change the soul.'

It was an idea both breathtaking and true. The most reliable way to change masses of selves, we've since learned, is by changing the ways by which they have to get along and get ahead. And so it was that the gamification of society triggered the 'Greed is Good' era, which represented a staggering transformation from the self of the anti-materialistic, communalistic hippies that had grown out of the mid-century's more collective economy. This new and intensified form of competitive individualism meant less support from employers and the state, which, in turn, meant more and more pressure placed upon the individual. To get along and get ahead, in this neoliberal world, meant being fitter, smarter and faster than your neighbours. It meant doubling and then tripling down on the fabulous power of Me.

Early signs of this revolution in self came in 1982 with the arrival of Esalen alumna Jane Fonda's *Workout Video*, which sold over a million copies and triggered a keep-fit craze that's never left us. The next year something strange started happening in America's maternity wards. As a study of over 300 million births starting in 1880 has since discovered, for generations, parents tended to choose traditional names for their children, such as John, Mary or Linda. But 1983 saw the beginning of a sharp increase in non-common names that would only become more pronounced in the 1990s and 2000s. According to one of the study's authors, Professor Jean Twenge, these mothers and fathers were seeking unique names because they wanted their child to 'stand out and be a star'.

It was into this manic phase of individualism that the self-esteem craze arrived. This was to be a rapturous copulation of the ideas of Ayn Rand, Esalen and the neoliberals, with each party falling upon the other as if in love at first sight. Personifying all of these forces was one remarkable man, the Esalen alumnus and powerful Californian Assemblyman John 'Vasco' Vasconcellos, who we last met at the Institute's disastrous 'Spiritual Tyranny' conference.

Vasco was an intense, brooding, scruffy man, prone to long troughs of depression, who had a large moustache and a larger temper and who often failed to live his message, that every individual was of infinite worth and should be treated as such. A Democrat by party allegiance, Vasco was also, in many ways, a classic neoliberal. He once complained that traditional liberals 'thought people couldn't take care of themselves' and even denied being one, 'in the old sense of thinking that big government with lots of money solves all the problems'. Declaring his belief in 'individuality, freedom and dignity – the American ethic!' he praised conservatives for being 'big on economic freedom' but chided them for their lack of commitment to the personal variety. He spent years attempting to institutionalize the Human Potential Movement by passing laws based on its ideas. The logo for one early attempt at doing so – his think tank 'Self-Determination' – was a stick man with 'I' for individual circled on its chest.

In 1986, Vasconcellos managed to persuade a sceptical California State Governor, the Republican George 'The Duke' Deukmejian, to fund a three-year task force that, with the help of Rand's former accomplice Nathaniel Branden and many others, would explore and promote the value of self-esteem. Vasconcellos was deeply influenced by his experiences at Esalen and was personally close to one of the fathers of Humanistic Psychology, Carl Rogers. He was an evangelist for the idea that the authentic core of the human animal was good. He believed that by convincing us all that we were special and wonderful, he could make us happier and more successful competitors in this new gamified economy. He argued that low self-esteem was the ultimate source of a huge array of social issues, including unemployment, educational failure, child abuse, domestic violence, homelessness and gang warfare – problems that cost the state of California alone more than ten billion dollars per year. He was convinced that raising the population's self-esteem would act as a 'social vaccine' and cure us of many of our failings.

Until the 1980s, self-esteem had been of interest mainly to educators, therapists and self-improvement-minded Californian-types. Vasco and his task force would be significantly responsible for making it an international pop-cultural product, whose effects are still felt today. But instead of saving us, or improving any of the social problems he promised, his plan backfired spectacularly.

The reason his project went wrong with such extraordinary effect was because there was, at its heart, a lie. The full story of Vasco and his task force has never been told. I spent a year digging it out, rummaging in California archives, reading thousands of letters, reports and documents, immersing myself in contemporary news reports and tracking down people who were there. What I discovered was a surprising tale of power, ambition, delusion, deceit and disastrous unintended consequences.

*

Little John Vasconcellos was just the most perfect little package of perfect. He was bound up so tight that you'd think he might just burst with all that godliness and humility and studious application. He was the first born of three, coming into the world at 01:00 on Wednesday 11 May 1932, and grew up to be an obedient Catholic, an altar boy, the smartest kid in class, whose mom swore he never, ever misbehaved. He adored his mother with such a clinging intensity that he aroused the ire of his father, who announced, on the day before the new term was to start, that he was to be sent away to a boarding school in San Jose. It was a terrible shock. Little Vasco cried and cried. He hated his dad for what he'd done. But he soon settled at the school, and excelled. He had ambitions to be a doctor until he saw a worm cut in half. The disgusting sight of both its ends wiggling put him off. One day, he decided to run for student office. But there was a problem. Little Vasco had no self-esteem.

As a devout Catholic boy, Vasco had always known that humans were broken and needed fixing. No matter how good his behaviour, he could only ever be a sinner. He was, in fact, so gripped by self-loathing that any positive notion about himself he might come across profoundly disturbed him. When his mother congratulated him on his academic successes he'd glare at her in a kind of furious shame. When he was older, in 1952, his sweetheart Nancy Lee wrote to him saying, 'I didn't realize how nearly perfect you are until I read your letter over again. The most impressive element is the feeling of humility you convey while describing your achievements.' He was uncomfortable with his naked body and so shy around others that when he ran for student office, and had to make campaign speeches, he could only stutter, 'Vote for me.' His lack of esteem ensured that he lost – and kept on losing. In the eighth grade, he ran for class president. 'I lost by one vote. Mine,' he said. He didn't vote for himself because, 'I'd been drilled never to use the word "I", never to think or speak well of myself.'

After studying law at the Jesuit University of Santa Clara, and a spell practising as a lawyer, Vasco entered politics. He was elected to the California State Assembly in 1966. But Vasco's self began failing. The incredible success he was achieving, at thirty-three, was at catastrophic odds with how he thought of himself. It was as if his brain couldn't contain the opposing forces, the world that was saying he was worth a great deal and the self that was telling him he was no better than that despised severed worm. At six foot three and over two hundred pounds he'd stalk the Capitol building in Sacramento, glowering and anxious, in his smart black suit, his perfect white shirt and arrow-straight tie, his hair cropped to military precision. This creaking state of contained intensity was becoming more and more unstable. It couldn't hold. He'd felt its first crack four years earlier, upon receiving a letter from his old college roommate that scolded him for his coldness and distance and accused him of not knowing how to love. But it was during that first year in office that he finally became a man in pieces. 'I found myself and my identity and my life coming utterly apart,' he said. 'My pain and confusion was so great that I had to go and seek help.'

That help came from an unusual, and distinctively Californian, Catholic priest. Father Leo Rock was a psychologist who'd trained under the father of humanistic psychology and encounter group pioneer Carl Rogers. 'For the next year, at least weekly, literally and figuratively, Leo held me – in unconditional positive regard,' Vasco recalled. 'He empathetically listened to me, and he assured me that I was OK, that it was OK for me to not know who I was and that I could trust that from within me would emerge some clues toward getting myself together again.' Through Rock, he began to glimpse a new view of life and the human animal that would eventually transform him. People weren't sinners, they were *amazing*. He read all the books he could find on humanistic psychology, eventually stacking up more than two hundred. 'My next step was to overcome my fears and to open up and become free and willing to experience my own feelings. That was utterly contrary to the way I was raised.' He attended his first encounter group at the home of the Hollywood actress Jennifer Jones, before continuing his journey inwards, towards personal authenticity, self-responsibility and self-love, at Esalen, where he participated in a series of eight workshops. He studied under Carl Rogers himself, who became such an important mentor to Vasco he'd one day describe him as 'almost my second father'.

The transformation Vasco's colleagues began to see in him, around the Capitol building in Sacramento, could hardly have been more dramatic. From being such a devoted Catholic that he thought the mass should never have been translated from the Latin, he became atheistic, unbuttoned and free. He grew his hair shaggy and curly, wore half-buttoned Hawaiian shirts on the floor of the Senate, a gold chain nestled in his chest hair. He drove around the city in a mustard Pontiac Grand convertible, its roof kept down even in the rain. One reporter described him as looking like 'a cross between a rock star and a drug smuggler'. He became a Human Potential evangelist, preaching the innate goodness of humans and handing out lengthy book lists to his political colleagues. He even managed to persuade a group of powerful politicians, including the House Speaker Willie Brown – a man so famously uptight he didn't even like people touching his clothes – to come to Esalen to experience encounter groups and soak in the hot tubs with the naked and the free. His miserable, cowering, self-hating Catholic self had finally washed away. In its place was a great, glowing letter 'I'.

Vasco had found his self-esteem. And he'd also found something else. Human Potential had granted him permission to be his true authentic self and that true authentic self, he quickly discovered, was *furious*. Colleagues would complain of being treated by him with undisguised contempt. He'd rant and roll his eyes and explode into states of wild wrath. 'I've seen John so mad he'd pretty nearly froth,' said one Republican Assemblyman. 'If one thing's out of place, he almost turns white with rage.' Another complained, 'It's all life and death with him. If you make one quip he can come back at you with a real moral zinger. You just can't kid with John. In committees he'll give out these great sighs. He tends to see hypocrisy or greed or a base attitude in another member's vote.' He'd become particularly enraged when he felt humiliated or that his ideas weren't being taken seriously. Eventually, he began losing his temper so frequently that other members were tasked with being by his side, during assembly sessions, to hold his hand and tell him everything was OK.

And it wasn't just his adversaries that suffered. The *Los Angeles Times* noted that, 'To an extent greater than in most other politicians' offices, his staffers speak negatively about their boss.' One of his secretaries wrote him a long personal letter begging him to treat his people with more kindness, and taking issue with his Esalen-esque stance that other people's upset was not his responsibility. 'It's fairly easy for you to justify not taking responsibility for anyone else's reactions, feelings etc in a particular situation but that's because it's easier not to listen or care,' she wrote. 'John, we're all human. We all have feelings.'

The 1970s found Vasco susceptible not only to fits of rage but to troughs of depression that would last for days, sometimes months at a time. He became tortured by an apocalyptic pessimism for his nation. 'Our culture is dying,' he wrote, 'if not indeed already dead. Evidence is manifold – rising rates of all social evils, drug abuse, family disintegration, crime, unemployment, taxes, inflation, alienation, prejudice, ecological rape, war, unconcern, apathy, crumbling institutions, liberation movements, fiscal bankruptcy of all levels of government and many non governmental institutions (evidencing their moral/human bankruptcy), growing alienation and lack of confidence by all kinds of people, in all kinds of institutions.' The problem, he decided, was that the people of America remained trapped under the old Christian delusion that humans were essentially rotten. 'Nothing explains it all except the single issue of how we envision ourselves as human beings. It is now demonstrable, though arguable, that the traditional, sinful, dualistic, self-denying vision of ourselves as human beings lies at the root of our problems.'

Things were bad. But there was a sliver of light. Vasco realized he was in an absolutely unique position. Not only did he have an idea about what could be done about all these problems, he also had the power to act. He could take everything he'd learned about the human potential and turn it into actual policy that would have real effects on thousands, perhaps millions, of lives. 'It is now time for a new vision of ourselves, of man, of human beingness, of human nature, and of human potential,' he wrote, 'a new theory of politics and institutions premised upon that vision.' He would make a sensational attempt, as one contemporary reporter put it, at arranging 'the unlikely marriage of Esalen and Sacramento'. There would be many false starts. It wouldn't be until the mid-1980s that he'd finally locate the mission that would ultimately define his life. Through the power of his office, he'd bless the population at large with the same transformation he'd been through. He would give the world self-esteem.

Vasco's dream took the form of a state-financed and mandated task force to promote self-esteem. To many in the 1980s, the notion that feeling good about yourself was the answer to all your problems sounded like a silly Californian fad. What Vasco could do was elevate it – stamp it with official affirmation – and then begin to fashion legislation around it, thereby institutionalizing it for the benefit of all. The task force could take testimony from people spotted about the west coast who already believed. They could investigate the causes of self-esteem and the forces that imperilled it. But perhaps best of all, they could recruit the world's finest researchers to give self-esteem the legitimacy of academia. They could prove, scientifically, that self-esteem was the vaccine for all social disease. But Vasco had a problem. He'd first have to conquer the Duke. California State Governor George 'The Duke' Deukmejian was a Republican with a reputation that out-no-nonsensed even his predecessor-but-one Ronald Reagan. What's more, he hated Vasco. And Vasco hated him. Not only did the Duke revel in his reputation as a tough fiscal conservative but California was on the verge of insolvency. Nobody thought it remotely possible that he'd release millions of tax dollars to indulge Vasco's ridiculous notion.

By the mid-1980s, Vasco's intellectual explorations had taken him to far shores, and he'd become notorious, around the Capitol, for some of his crazier ideas. He wondered, for example, if a special form of 'gentle birthing' might lead to less violent humans. He was apparently at least curious about the idea that children having sex with their parents was not, in fact, harmful to the child but natural and healthy; he'd read reports of women expressing pride at having lost their virginities to their fathers and invited a proponent of paedophilia to join his 'network on sexuality'. It was just these kinds of nutty notions that made him a natural enemy of the Duke. But Vasco had at least one thing going for him. In 1980 he'd been appointed Chairman of the Ways and Means Committee. Not only was he in charge of all the money, now, there was arguably only one person in the Capitol in a position of greater power. It was Vasco's misfortune that that man was the Duke.

He made his first attempt at having his task force mandated into law in 1984. He managed to push it successfully through the Assembly and was about to take it to the State Senate when he suffered a massive heart attack. His belief in the power of positive thinking was such that, in an attempt at curing himself, he wrote to constituents asking them to imagine themselves with tiny brushes swimming through his arteries, scrubbing at the cholesterol. 'Focus yourself, your attention and energies on me, my heart and my healing. Picture my constrictions.' They were to do this whilst singing, to the tune of 'Row, Row Your Boat', 'Now let's swim ourselves up and down my streams/Touch and rub and warm and melt the plaque that blocks my streams.'

It didn't work. As the vote in the Senate was taking place, Vasco found himself in a bed in Menlo Park recovering from seven-way coronary bypass surgery. Unable to personally shepherd in all the votes, his dream failed. It was a time of anguish and blackness through which he was helped by Carl Rogers, who, following Vasco's release from hospital, treated him to a seafood buffet at his favourite La Jolla restaurant then took him home, where the great psychologist listened to his tales of loneliness and depression.

Following his heart attack, Vasco did appear to be somewhat better at managing his temper. Although he put this improvement down to his self-work, friends pinned it on the lifestyle changes that followed illness, not least his giving up sugar (he used to 'sit in committee and eat 1100 cookies,' said one). He'd also cut his work hours and taken up racquetball. But, even then, he'd frequently revert to being the same old John. A colleague from the Ways and Means Committee told the *Los Angeles Times*, 'You walk around here on eggshells. "Is John in a good mood today or a bad mood?"' His best friend Mitch Saunders said, 'I think there are lots of times when John really doesn't know how to treat people with the values he espouses.' A gossipy item in the *Sacramento Bee* reported his losing his composure when a microphone failed him: Vasco apparently wanted to 'honour, inspire and empower every Californian – except those who make him angry'.

His next attempt at getting his bill enacted into law passed through both houses. But then the Duke struck, personally vetoing it. 'So I got shrewd,' said Vasco. He decided to rename his project the 'Task Force to Promote Self-Esteem and Personal and Social Responsibility'. 'Right away I got the attention, and the support, of the traditionalists,' he said. He also asked for less money, reducing the proposed budget from $750,000 a year to $735,000 over three. Then he steeled himself for a series of intense conversations with the Duke himself.

He made a powerful argument. Self-esteem would transform people into better players in the new competitive economy. 'Healthy self-esteeming' citizens would be 'responsible, productive, creative and satisfied workers not absent from the job' and 'make America competitive again'. Moreover, just by raising self-esteem, you could help solve a massive array of chronic social problems: child abuse, educational failure, teenage pregnancy, alcohol and drug abuse, welfare dependency and crime and violence. These kinds of problems cost the state of California billions of dollars every year. And self-esteem wasn't just some silly Esalen fantasy. Part of the task force's job would be to bring it out of the hot tubs and expose it to serious academic rigour. They'd recruit the finest experts and add their scientific data to the truth that self-esteem was the secret cause of human thriving. And Vasco had little doubt that that was what they'd conclude. He even wrote it into the bill itself, which asserted that, 'A body of research studies now exists which documents the causal relationship between self-esteem, which is a developed sense of one's inherent worth as a person, and the growth and development of healthy, responsible individuals.'

His breakthrough came on their third meeting. The Duke listened, yet again, to his speech. 'I know self-esteem is important,' he conceded. 'But why should the government get involved with doing it?' Vasco sensed his moment. 'First, Governor, there's so much at stake here, we can't afford to have it hidden in some university. We need to involve the entire California public. By spending a few taxpayer dollars, we can collect the information and get it out. If that helps even a few persons appreciate self-esteem and how they can live their lives and raise their kids better, so we have less welfare and crime and violence and drugs, that's a very conservative use of taxpayers' money.' The Duke's expression immediately lifted. 'I've never thought of it that way before.' On 23 September 1986, Assembly Bill AB3659 was signed into law.

The response from the media was immediate and savage. An editorial in the *San Francisco Chronicle* called Vasco's task force 'naive and absurd' and struggled to understand what the Duke had been thinking. 'Big John is very influential in Sacramento and his colleagues have indulged his odd ideas for a simple reason. He's in charge of the money box key and that creates a fondness among his colleagues, both for the man and the vagaries he dreams up. That does not explain, of course, why George Deukmejian, the tight-fisted protector of the public purse, suffered a momentary loss of judgement and signed Vasconcello's pipe dream nonsense. He is on a path that will lead to loss of self-esteem.'

Nothing made Vasco more angry than his ideas not being taken seriously. He was about to become the joke of America.

*

Until Monday 9 February 1987, California's Self Esteem Task Force was largely state news. But on that morning, the nationally syndicated cartoonist Garry Trudeau, who'd apparently been tickled by the idea of Vasco's crusade, began an extraordinary two-week run of his popular *Doonesbury* strips devoted to it. It starred a new character, young Barbara 'Boopsie' Ann Boopstein, a twenty-five-year-old LA actress and spiritual medium who channelled a 213,555-year-old warrior named Hunk-Ra who'd been invited onto the Task Force on the basis of her 'twenty years of feeling good about myself and out of body experiences'. The effect of the national, high-profile publicity was immediate. By the end of that day, reporters from press and television were crowding Vasco on the floor of the Assembly Chamber. But they didn't appear to be treating his Task Force with the gravitas he might have hoped. When one journalist asked him about his letter asking constituents to clean his arteries with tiny brushes, Vasco responded defensively: 'I've been willing to break new ground, to be more immediate and to publicly show myself and live my beliefs.' The scrum became such that the Speaker was forced to angrily order them to cease and desist so the politicians could conclude their business.

And that was just the start of it. Overnight, everyone in America seemed to be talking about Vasco and his task force. Sadly, little of that talk was kind. Stand-up comics joked about it and Johnny Carson cracked wise on his *Tonight Show* in front of millions. Rival politicians gave dismissive briefings – 'You could buy the Bible for $2.50 and do better' – whilst newspapers sneered. The *San Francisco Examiner* called the idea 'laughable'. The *Pittsburgh Post Despatch* wrote, 'California is the state that produced Jerry Brown, the People's Temple, Sister Boom-Boom, drive-in churches, Charles Manson, the Esalen Institute – and a governmental Task Force to Promote Self Esteem . . . Now there's one more California joke to tell at cocktail parties around the nation.' The *Wall Street Journal*'s story bore the headline 'MAYBE FOLKS WOULD FEEL BETTER IF THEY GOT TO SPLIT THE $735,000.' Earlier, the *New York Times* had dismissed it as another California curio, the latest in a 'number of improbable ideas that might have been rejected elsewhere [but] have easily taken root here.' And now Trudeau's cartoons had, as one Los Angeles newspaper acknowledged, turned Vasco's beloved project into a 'national joke'.

Vasco was livid. In room 6026 of the Capitol building in Sacramento he'd sit at his desk in his dark, rumpled suit, playing with a Day-Glo orange Slinky, fuming, 'I'm sick of it.' His office was lined with photos of Martin Luther King, quotes from Malcolm X and George Bernard Shaw and with books –

I'm OK, You're OK; *On Becoming a Person* – and pictures. There was a photo of the buttoned-up pre-enlightened Vasco shaking hands with JFK, and a line drawing of an animal that'd been cut in half and was peering inside its body, smiling at the marvel of itself. He kept a collection of Skippy peanut butter and a menagerie of toys, including a large teddy that wore a T-shirt saying 'SELF ESTEEM'. This task force had been the mission of his life. What was happening was an insult and a calamity. 'I'm sick of the scornful stuff that has been dumped on this,' he complained. 'It's bizarre to me that someone would not take this group seriously.' His mother had been personally upset at the 'cocktail party joke' article and Vasco railed against 'folks who aren't able to stand on their own and need to destroy anyone who is . . . The cynics who find that it's smart to make wisecracks and say funny things, cartoonists or otherwise.' The media, he complained, were 'terrible, cynical, skeptical and cheap'. Their problem? 'Low self-esteem.'

Even the machinery of government appeared to be treating his project with contempt. The task force, which was due to run for three years, had been assigned a former smoking closet that was now being used as a photocopy room to work from. 'This has been really difficult and painful for me,' wrote the group's Interim Director, Dick Vittitow, in a private memo to the task force chairman, Andrew Mecca. 'It is the present Xerox and paper supply room in the basement. It has a pillar, three foot square, in the centre'. Observing that 'space in Sacramento is power', Vittitow refused to accept it. 'It would not only be inadequate for the hard, sophisticated, terribly demanding work of the TF, it would be depressing, and extreme in its impact on their self-esteem.' But there was good news too. Something remarkable seemed to be happening. The response from the *people* of California had been great.

Between its announcement and the task force's first public meeting in March 1987, the office received more than two thousand calls and letters and almost four hundred applications to volunteer – a figure that broke state records. Fan mail outnumbered complaints by ten to one. More than three hundred people came forward to speak at the public hearings that would be held across the state. And even if the tone of the media hadn't always been respectful, it was still true that Vasco himself was now a national figure. In the weeks following Trudeau's cartoons, he'd appeared everywhere from *The Economist* to *Newsweek* to the CBS *Morning Show* to the BBC. The *Los Angeles Times* and *San Francisco Chronicle* were planning major stories on him. This, he sensed, could be a huge opportunity. 'I've gotten more attention in the last several weeks than in twenty years,' he said. 'The purpose of the task force was to elevate self-esteem into public awareness throughout California. Trudeau has made it nationally recognized. Now I think we have a real chance to make some history.'

But if he'd ever be able to do that, he'd first have to find a way to wrench the media conversation upwards. And things, on that front, were going from unfortunate to ridiculous. It began with the grand announcement of the task force's twenty-five members. On the upside, it was a diverse group including women, men, people of colour, gays, straights, Republicans, Democrats and a former police officer and Vietnam veteran who'd been awarded two Purple Hearts. But on the downside it also included a white man in a turban who predicted the work of the task force would be so powerful it would cause the sun to rise in the west, and a female psychologist who argued that low self-esteem was positively correlated with being raped. A delighted writer from the *Los Angeles Daily News* declared the reality of the proceedings to be 'even better than the stuff in Doonesbury'. The *Los Angeles Herald*, meanwhile, could barely contain itself as it reported their public debut. On a front-page story, it told how the famous family therapist and Esalen chief Virginia Satir had asked her fellow members to close their eyes and imagine a 'self-esteem maintenance kit' of magic hats, wands and amulets. 'Task force members then spent the next fifty minutes attempting to find ways to cut down the media guffaws.'

By now, the task force had begun hearing testimony from believers up and down California. They heard from an LA deputy sheriff who toured schools preventing drug abuse by raising esteem. 'We're teaching students right off the bat, "You are special. You're a wonderful individual. You are very special."' They heard from masked members of the Bloods and the Crips, one of whom pinned his criminality on a thirst for self-esteem. 'My father was quick to whip me on the butt when I did wrong, and wasn't so quick to pat me on the back,' he said, his eyes peering out from above the handkerchief that was tied around his face. 'We had more one-on-one discussions when I did wrong, so I felt I had to do wrong to get his attention.' One school principal urged the task force to recommend having elementary pupils increase their sense of self-importance by carrying out evaluations on their teachers. Another told the panel, 'I love you, I really do. That's what I tell my students because each of them is a special person.'

Helice Bridges was an expert witness who'd written to Vasconcellos when the task force was announced, almost incontinent with enthusiasm for the man and his project. 'When we spoke the other day, I had the experience that Christopher Columbus had just arrived in America,' she wrote. 'I don't know whether to congratulate you for your persistence, dedication and keeping your vision on course or to simply say "I love you!"' Introducing herself to the panel ('I'm known today as the Blue Ribbon Lady, but a lot of people call me Sparky,') she explained how she'd dedicated her life to distributing hundreds of thousands of blue ribbons that said, 'Who I Am Makes a Difference'. 'Some time ago, I just walked around and decided I was going to tell as many people as I saw how great they were,' she said. 'What I noticed was that people, wherever I went, started to cry. They said, "My God, that's the first time in my life I ever received recognition."' Helice awarded the panel their own ribbons and instructed them how to use them. 'Hold the blue ribbon, say your name, then say, "I have a blue ribbon and it says, 'Who I am makes a difference.'" Tell yourself how great you are. Then you say, "Do I have permission to put it on me?" Heck yeah! Stick it on yourself! Just above your heart! Up toward your dreams coming true! Know that the ribbon is magic. Whenever you see it, you'll always have great thoughts about yourself and every other human being in the world.' Vasco was sufficiently impressed by Sparky that, after the task force's completion, he picked her out as one of his highlights.

Other contributions were more lucid, not least those of a famous self-esteem veteran and Ayn Rand ally who turned out to be a major influence on the project. Nathaniel Branden, according to task force Executive Director, Bob Ball, 'made significant contributions to our task. We had lots of contact with him. Numerous meetings.' Much of Branden's efforts were engaged with helping them define self-esteem's causes and threats. It comes, he advised, by 'encouraging the child to be in love with his/her own existence'. Their ambition should be to 'create a world in which people understand that to honour the self is to practise selfishness in the highest, noblest and least understood sense of this word.'

Another celebrated ally turned out to be self-crowned 'Emperor of Esalen' and charismatic encounter-group guru, Will Schutz. He attacked the 'cascade of invective and ridicule' that had been directed at the group and said, 'From my vantage point, working in the area of human behaviour for over thirty-three years, John Vasconcellos and the Self-Esteem task force are exactly right. Self-esteem is, indeed, the heart of the matter.' By the late eighties, Schutz had morphed into a self-esteem business consultant, promising blue-chip clients including Procter & Gamble, the United States Army and General Motors that boosting the self-esteem of their workforce could increase productivity by up to 300 per cent.

Not everyone who contacted the task force was so supportive. Janet Mayfield from Orange County told the panel, 'I don't think the problem is lack of self-esteem. It's selfishness. If you observe any child that's born into the world, he cries and screams and has absolutely no regard for his parents' sleep. From the very beginning we are self-centred.' The office also received an angry letter from the singer Randy Sparks whose request for a position had been snubbed, a decision he considered an opportunity wasted. 'Two weeks ago my band and I played to thousands at The Stockton Asparagus Festival,' he wrote. 'I feel I've been insulted, and I want you to be aware that when members of my audience suggest I'd be the perfect addition to your team of self-esteem volunteers (and this has already happened) I'm under no obligation to remain silent.' In 1988, Sparks released his album *Oh Yes, I'm A Wonderful Person and other Musical Adventures for those of us in search of Greater Self-Esteem.* It included the numbers 'Oh Yes, I'm A Wonderful Person', 'I'm Good', 'I Like Me', and 'Oh Yes, I'm A Wonderful Person (Reprise)'.

With the national media being given so much to snigger over, it was beginning to look as if there was no way to rescue Vasco's mission. But he had one more move to play. He'd promised the legislature that the task force would compile the world's best evidence that self-esteem really was a 'social vaccine' that could make us fitter, happier and more productive. And, on that front, there had been some good news. The University of California system had not only agreed to recruit seven of their professors to supply this data, they'd decided to publish it all in a book in their prestigious University of California Press imprint. This, it was agreed, was a significant coup. The task force chairman, Andrew Mecca, said the UC professors gave them 'enormous credibility and a base to operate from'. Executive Director Bob Ball declared their decision to publish one worth 'rejoicing'. He promised the public it would be 'revolutionary research' that would be of 'historic importance'.

But before they could take full advantage of the public relations opportunity all this afforded them, there was just one formality to get through. The professors had to actually do their work. That they would endorse Vasco's belief was of critical importance. He'd now earned notoriety throughout America and beyond for his extreme convictions. He'd been criticized and mocked for them and that was exactly the kind of humiliation that tormented him most. Even his beloved mother had been upset by the attacks. He'd promised the Duke and the legislature that the 'causal relationship' was backed up by research. That same bill contained a clause that said his project could be cancelled, its funding pulled.

It would be at 7:30 p.m. on 8 September 1988 at the El Rancho Inn in Millbrae that an expectant Vasco and his task force would get their news. For Vasco, it would be personal. If the professors, for some lunatic reason, decided he was wrong, it could be a calamity.

*

The star of the day on which everything hinged was University of California academic Dr Neil Smelser. It was Smelser who'd been responsible for coordinating the work. He was an undeniably serious individual – an emeritus professor of sociology at Berkeley who'd pioneered the field of economic sociology. News of what took place, that evening at the El Rancho Inn, was released to the public in January 1989. 'Through a contract negotiated between the Task Force and the University of California, one of the world's most prestigious research institutions, seven distinguished professors undertook an examination of "the world's most credible and contemporary research" in each of the areas of specified concern,' they said. 'What did they find? In the words of an eighth professor, Dr Neil Smelser, "The correlational findings are very positive and compelling."

The significance of these research summaries is reflected in the decision of the University of California Press, a premier publisher of scholarly research, to accept the papers for publication.'

The effect of this good news was immediate. The Associated Press wrote a glowing wire piece confirming that 'University of California research' and testimony were 'lending legitimacy to the commission's founding premise: Poor self-esteem is closely linked with alcoholism, drug abuse, crime and violence, child abuse, teenage pregnancy, prostitution, chronic welfare dependency and failure of children to learn.' The headlines quickly piled up: 'Self-Esteem panel finally being taken seriously'; 'Commission on Self-Esteem Finally Getting Some Respect'; 'Boopsie Board Gains Respect'. Speaking to reporters, Bob Ball acknowledged happily, 'I think we're now gaining a great deal of credibility.' The Duke was so impressed by the professors' work that he sent copies to his fellow governors, saying, 'I'm convinced that these studies lay the foundation for a new day in American problem solving.' The task force held their final public meeting in April of that year. The mood was celebratory. Students from Lincoln Elementary School in Long Beach, who'd been receiving self-esteem lessons, appeared as special guests, singing songs and reciting poems and proudly cheering themselves. 'There is magic in me because I believe in myself,' they chanted. 'We're kids! We're super-cool! We're great!'

It was an extraordinary turnaround captained by the man that task force Executive Director Bob Ball rightly called 'a master politician'. All Vasco's team had to do now was build upon this tone for the grand publication of their final report in January 1990. Having rejected titles that ranged from the pompous ('Of Human Worth') to the imbecilic ('Save Your Ass: If You Want To, You'd Better Read This') they settled on Vasco's own pleasing suggestion – 'Toward a State of Esteem'. Pumping up the now largely sympathetic media, the politician promised it would be 'a giant step for humankind' and that it showed how 'self-esteem is the social vaccine'. Bob Ball declared, 'We're now far more certain of the connection between self-esteem and the problems we face. All of the evidence turned up that the connection is even more important than we thought.' The report was trumpeted by a press release that said, 'The relationship between poor self-esteem and society's problems has definitely been confirmed.'

Toward a State of Esteem went on to be a victory that lay far outside the reasonable hopes of anyone who'd witnessed its humiliating origins. The Governor of Arkansas, Bill Clinton, who'd privately mocked Vasco and his project, now publicly endorsed it, as did serious figures including Barbara Bush and Colin Powell. Although, naturally, it could never have claimed to have convinced every member of the American press, the majority bent knees towards it. The *San Francisco Examiner* said the task force, which was 'once the butt of "only-in-California" jokes, has issued its final report – and nobody's laughing.' The *Philadelphia Enquirer*: 'It looks like John Vasconcellos may have the last laugh. And he feels good about that.' *Time Magazine*: 'The sneers are turning to cheers.' The *Ledger*: 'There is a direct link between low self-esteem and the ills of society.' The *Daily Republic*: 'OFFICIAL: SELF-ESTEEM IS A "SOCIAL VACCINE"'. The *Washington Post*: 'Three years after its birth amid late night television gibes, Gary Trudeau cartoons and other negative vibrations, California's official Self-Esteem Task Force unveiled its final thoughts on the condition of mankind today in a 144-page report as dry and academic as any of its genre. Yet that seemed, in a way, to be a victory. For the report itself and a look at the way the nation's largest state is doing its business suggest that the self-esteem movement has now moved directly into the mainstream, with school administrators, welfare caseworkers and correctional officers experimenting with self-motivating techniques once thought appropriate only for long weekends at Big Sur.'

Self-esteem had been legitimized, now, by science and by government. It began to intoxicate America and then the world. The man they were calling 'The Johnny Appleseed of Self-Esteem' appeared on shows such as *CBS Morning News, The Today Show* and *Nightline* and on the BBC, German national radio and Australia's ABC. He was invited to speak in Norway, Japan and at the Supreme Soviet Parliament in the USSR. The report went into reprint in its debut week and went on to sell an extraordinary sixty thousand copies, smashing all state records. Three months later, the *LA Times* reported that the self-esteem movement was now ready to 'break out across the state and country, if not throughout the world'. Vasco's publicists approached *The Oprah Winfrey Show*, persuading her producers to speak directly with him about covering his precious subject. On 15 June the *Baltimore Sun* reported the next breathless development: 'SELF-ESTEEM HAS ARRIVED, SAYS OPRAH WINFREY. Schools may teach it, politicians praise it, but when Oprah Winfrey does a prime-time special on it, a trend has arrived.' The show would examine why Oprah believed self-esteem was going to be one of the 'catch-all phrases for the 1990s'. Interviewed were Maya Angelou, Drew Barrymore and John Vasconcellos.

About that, Oprah was right. Four months after the launch of *Toward a State of Esteem*, the papers were reporting that self-esteem was 'sweeping through California's public schools' with 86 per cent of the state's elementary school districts and 83 per cent of high school districts implementing self-esteem programmes. In Sacramento, students began meeting twice a week to decide how to discipline other students; in Simi Valley, kids were taught 'It doesn't matter what you do, but who you are.' Fifty National Council for Self Esteem chapters were formed across the US, and eleven more established internationally. Political leaders from Arkansas to Hawaii to Mississippi began considering their own task forces. When Maryland launched theirs, 'phones were totally blocked . . . people were calling and mailing, begging, begging . . . to get on this task force'.

As the months became years, the self-love movement only became more engorged. Everybody had to feel good. Defendants in drug trials were rewarded with special key chains for appearing in court, whilst those who completed treatment were given applause and donuts. Television evangelists preached, 'People who do not love themselves can't believe in God!'; kindergarten five-year-olds were issued T-shirts saying, 'I'm loveable and capable'; children were awarded sports trophies just for turning up; a Massachusetts school district ordered children in gym classes to do skipping rope without actual ropes lest they suffer the self-esteem catastrophe of tripping; grades were inflated ('It's not grade inflation, it's grade encouragement,' explained one teacher). A 1992 Gallup poll found 89 per cent of Americans believed self-esteem to be a 'very important' factor in motivating a person to work hard and succeed. Meanwhile, police in Michigan seeking a serial rapist instructed the public to look out for a thirty-something male with medium build and 'low self-esteem'.

Crucially, as Vasco noted with pleasure, the self-esteem doctrine had also begun trickling into law. The sociologist Professor James Nolan Jnr writes, 'A look at state law in all fifty states reveals that Vasconcellos's assessment of self-esteem's widespread acceptance and institutionalisation is on the mark. By the middle of 1994, some thirty states had enacted a total of over 170 statutes that in some fashion sought to promote, protect, or enhance the self-esteem of Americans. The majority of these (around 75) are in the area of education.' Schools in Britain, too, became infected. Educational psychologist Dr Laura Warren, who taught in the 1990s, recalled this period well, remembering her school's edict for mauve pens to mark errors, in place of the negative red. 'It was a policy of "reward everything that they do",' she told me. 'That turned out to be a terribly bad idea. Very harmful. Of course, it came from America.'

By opening the lid of self-esteem and peering in, you can easily spot Ancient Greek ideas of the human as a separate node of importance whose duty is self-improvement; you can see the Christian ideas of the struggle with the inner self; you can see the Mind Cure movement, which brought with it the miraculous healing power of thoughts. And then there's Carl Rogers, with his belief that the core of the human is good, and Esalen, that hot zone of the Human Potential Movement, which preached that people were as gods and had to be open and true to themselves and take responsibility for everything that happened to them.

It's also no coincidence that when trying to get his task force approved, Vasconcellos made an economic pitch to the Duke. In this tough new neoliberal era, the comforts of the Great Compression were gone. Increasingly, the state was not there to look after you any more, protecting your income, respecting your union, catching your fall. To get along and get ahead, in this competitive age, you had to be fit, ambitious, ruthless, relentless. You had to believe in yourself. Self-esteem was a simple hack that would make you a fitter, better, more winning contestant in the neoliberal game, and who wouldn't want that? The task force was able to succeed so fantastically because it was an idea people were ready for. It made sense, in some deep, unspoken way, of the economic reality from which their sense of self was now emerging. It wasn't merely that they wanted it to be true, it *felt* true.

But it wasn't.

The 1990s self-esteem boom was nourished by Vasco's task force whose credibility turned largely on a single fact: that, in 1988, the esteemed professors of the University of California had analysed the data and confirmed his hunch. The only problem was, they hadn't. The scientists had not found self-esteem to be a social vaccine. When I tracked down one renegade task-force member, he described what took place, following the crucial Professor Neil Smelser meeting of September 1988, as 'a fucking lie'.

*

Self-esteem is where the story, for me, gets personal. In a book by sociologist Professor John Hewitt, I read that people like me who embark on quests for higher self-esteem are just characters blindly acting out our culture – protagonists in a classic, but foolish and imagined, hero plot. 'In this myth of self-esteem,' he writes, 'we have a profound mirror of our culture, a way of gaining some perspective on the way we think and act. It is not a story of ancient heroes and military triumphs but a contemporary tale in which men and women overcome mainly psychological obstacles to success and happiness. Its heroes are not soldiers but positive thinkers who lift themselves up by their psychic bootstraps; its priests and preachers are psychologists and therapists.'

When I spoke to John, I told him there was something about the idea of being able to solve our problems just by raising self-esteem that seemed inherently American.

'It is,' he said. 'It's part of our cultural mythology. It's American exceptionalism. Our culture says we've created a new kind of human being and we're free.'

'As in, free to be whoever we want to be?'

'Exactly.'

'But it's a false idea?'

'Oh yes,' he said. 'It's a myth. A social construction.'

Between 1988 and 1995, the period in which the movement was at its peak of influence, I was a teenager. I spent my formative years being told by teachers, therapists and well-meaning friends that my problems were rooted in low self-esteem and solving them was simply a matter of raising it. I believed what they told me, and spent years trying. Nothing worked. It was only when I started this research that I discovered that I'd been wasting my time. That 'golden city on top of a hill' I'd imagined – the place that, when I reached it, would magically transform me into the perfect version of myself – was a mirage. I could hardly believe it. My fight with low self-esteem was *who I was*. This experience, more than any other, made it viscerally clear how much our sense of who we are is a construction, and how much of that construction is built from the lives of others – many of whom we've never heard of, are long dead and who, it turns out, were wrong.

So what was I going to do now? I'd sat there, in all those therapy rooms, confidently pointing the finger at my strained and chilly childhood, which was macerated in all that self-hating Catholicism, and decided *that* was the cause of how I felt about myself, the driver of my 'neurotic perfectionism'. Then, I'd accepted that my grumpy, antisocial, neurotic side couldn't possibly *define* my self because the self is an 'illusion'. The ease at which it seemed to vanish, on my final day at Esalen, seemed to confirm that – even if I did slip up a bit at the end. All that unhappiness, I'd been led to believe, had been triggered by a fixable bug in the software.

Now, not only had I been robbed of a fix, I'd been left with the nightmare conclusion that no consumer of therapy ever wants to face: what if my self-loathing was actually an entirely rational reaction? What if I was deluding myself, dismissing this bad aspect of my character as an illusion? Perhaps I felt bad about myself because I *ought* to feel bad about myself. Maybe that cunt, which I'd brought forth at Esalen with such glee, was really me.

*

In an attempt to find out how America, and then the world, got conned so spectacularly by Vasco and his people, I travelled to Del Mar, California, to meet the white man in the turban who'd predicted the task force's work would cause the sun to rise in the west. David Shannahoff-Khalsa was a kundalini yoga practitioner who believed meditation to be an 'ancient technology of the mind'. Having devoted his life to using neuroscience to see if yogic breathing could confer incredible health benefits, he'd been infuriated when the task force failed to include his findings in their conclusions. But something had been vexing David even more than that. In fact, he'd been so disillusioned by the task force's final report, he'd refused to sign it. When it was published, above his name in its opening pages, there was but a blank space.

On a sunny morning in March, I knocked on the door of his bungalow. He welcomed me in, looking hardly different to how I'd seen him in old task-force photographs: face narrow, eyes sharp, turban blue. In his backyard there was an old racing bicycle with its rear wheel removed, presumably for use as an exercise machine; on his bookshelves bestsellers by Norman Doidge and the DSM-IV; on his walls picture after picture of the dogs he's owned during his life, all of whom have been golden retrievers and two of whom have been called Bubba. 'Stop that!' he shouted to the latest incarnation of Bubba, who bounded around me with grovelling excitement. 'Show some self-esteem!' He presented me with a selection of cheeses to nibble on before going over to deal with the dog. 'You're being tied up because you're not being good,' he told Bubba affectionately before settling next to me at the living-room table. 'He's overly social and very spoiled and very loved. You know he doesn't eat dog food. He's never eaten it.'

'What does he eat?' I asked.

'Human food,' said David. 'Lay down, Bubba! He's about five hundred dollars a month in food.'

The dog panted at me in hopeful frustration. 'You're lucky, Bubba,' I told him.

'He is lucky,' said David. 'But I'm lucky to have him. Feel free to ask any questions you want. I'll be candid with you. Lay down!'

Next to David, on the table, was a shiny, thick, red hardcover book: *The Social Importance of Self-Esteem*. This was the collected work of the University of California professors, with introductions by Vasco and task force chairman Andrew Mecca and a summary of the content within that had been written by Dr Neil Smelser, the professor who'd coordinated the academic endeavour. It had been published, by the university, in July 1989. David opened the book with careful purpose. 'Let me show you, young man.' He flicked through, settling eventually on page fifteen – a section from Smelser's summary of the professors' findings – and scanned it, squinting at the type. 'Let's see.' He began reading: '"The news most consistently reported, however, is that the association between self-esteem and its expected consequences are mixed, insignificant or absent."'

This was Smelser's real view of the science. And it was, of course, a radically different conclusion to that which had been pushed so successfully to the media and public by the task force. David claimed to have been in the room when Vasco first laid eyes on the preliminary drafts of the professors' work. 'We're sitting at a table, I believe it was in San Francisco or Sacramento, and John Vasconcellos is sitting here and I'm sitting over here, one or two people away from him,' he said. 'I remember him going through them and he looks up and he says, "You know if the legislature finds out what's in these reports they could cut the funding to the task force." And then all of that stuff started to get brushed under the table.'

'And how did they do that?' I asked.

'They tried to hide it. They published a [positive] report before this one,' he said, tapping the red book – which went on to sell fewer than four thousand copies, nowhere near the volume of the task force's own much more positive final report, *Toward a State of Esteem*, which shifted sixty thousand and which, said David, deliberately 'ignored and covered up' the science.

'So that was a *dishonest* effort to cover up what the scientists were saying?'

'Oh, it was absolutely dishonest,' he said.

As we talked and nibbled cheese, David revealed that Vasco had also lied about his life story. He'd sometimes been vague about the cause of his personal breakdown in 1966, but around the time of the task force it suddenly coalesced into a convenient tale of crisis, struggle and victory over low self-esteem. But this, it seems, was not true. 'I don't know at what stage in his life he realized he was homosexual,' said David, 'but that was the core problem. He was raised Catholic and had the usual sorts of Catholic guilt. That was probably the kernel of his personal crisis.'

'He was gay?' I said.

'Right. And we all discovered it. Some, obviously, who knew him well knew it right away. And then it sort of came out. It wasn't an issue for us.' Vasco died in 2014, aged eighty-two, a decade after he'd quit politics for a position as scholar-in-residence at Esalen. Having had this claim independently verified, I thought it a rather melancholy discovery, that this lifelong evangelist for personal pride and authenticity apparently never felt able to be public with his sexuality.

Our talk returned to the self-esteem lie. It was in September 1988 that Dr Smelser presented the academics' conclusions to the task force. The media and public were told he'd been purely positive and were given just one direct quote: 'The correlational findings are very positive and compelling.' But this is what I found confusing. It was also Smelser, of course, who had written the damning 'mixed, insignificant or absent' line in the book. So which one was it? What did Smelser really believe? And what had *actually* happened at that meeting? What had Smelser told them about the scientists' findings on low self-esteem? I'd been trying to find out, but days in the state archives in Sacramento had turned up nothing.

But then, quite unexpectedly, I caught an enormous break. In despair, I'd ordered up a selection of old audio cassettes – an incomplete and apparently random record of public presentations that also seemed to include a handful of meetings. On one of them, I saw the words: *September 8th, Millbrae.* My heart leapt. That was it. As I approached the man at the archives counter, I had to force myself not to run. I asked for a tape player and a set of headphones. Then I found a quiet corner and hurriedly slotted it in.

*

The sound was hissy and faint and, to make out the voices, I had to press the headphones into my ears and screw my eyes shut. What I eventually heard, though, was clear enough: it was Dr Smelser's presentation. And it had been nowhere near as upbeat as the task force had claimed. I listened as he announced the professors' work to be complete but worryingly mixed. He talked through a few areas, such as academic achievement, and said, 'These correlational findings are really pretty positive, pretty compelling.' This was, surely, the quote the task force used. They'd sexed it up for the public.

But, much worse than that, they'd omitted what he'd said next: 'But in other areas the correlations don't seem to be so great and we're not quite sure why. And we're not sure, when we have correlations, what the causes might be. Let's take, for example, one of the areas where the findings are a little bit loose, which has to do with self-esteem and alcoholic abuse. By and large there are positive correlations here, but what does that mean in terms of cause? Do these people go to drinking because of an earlier history of self-doubt, self-degradation, worthlessness and so on? Or is it the other way around? Does the involvement in alcohol for years or decades constitute the causal basis for the feelings of worthlessness that we discovered in people who have been involved in that?'

This was the crux. Vasco had promised the legislature the data showed the 'causative' effects of low self-esteem. And this was precisely what Professor Smelser was saying was not there.

Correlative findings can be as good as useless; as every science student knows, correlation does not equal causation. You might find domestic violence to be correlated with liking Dolly Parton (you might find that, I have no idea) but that doesn't mean Dolly Parton is the cause of domestic violence or, indeed, that eradicating Dolly Parton would be a vaccine for it. At the end of his presentation, Smelser gave the task force a warning. The data, he said, was not going to give them something they could 'hand on a platter to the legislature and say, "This is what you've got to do and you're going to expect the following kind of results." That is another sin,' he said. 'It's the sin of overselling. And nobody can want to do that. You don't want to do that. Certainly, we don't want to do that.'

David had told me that, when he realized the task force were distorting Smelser's presentation so egregiously in the media, he'd made a protest at a meeting attended by Vasco. 'I'd always speak my mind,' he said, 'because I felt it was my responsibility to say not what John Vasconcellos wanted me to say, or Bob Ball, our executive director, or Andy Mecca, our chairman.' Later, I found another tape that actually captured his moment of insurrection. This gathering was on 2 February 1989, shortly after the media had been supplied the lie. The first person to speak up was, in fact, Wilbur Brantley, the Republican member who'd served with honour in Vietnam. 'There's a need for us to get to the marrow of the bone but I doubt, strongly, that that's going to be either in our final report or any other piece of work,' he said. 'People like Nathaniel Branden have been in this business over thirty years and he's not yet found that marrow. I'm very much concerned about the things we're saying in the media, and what's been printed in the media. We're implying we've found that marrow. And we have not. We're implying that there's an agreement among the professors who are producing the research. It's not true. End of speech.'

Then, following an unrelated discussion, David took his turn. He told the task force that, when he'd read the official account of the meeting, he'd thought, 'I don't think that was the bottom line that Neil Smelser would have wanted to leave with us.' So he'd called Smelser up. 'I asked him what he thought about it, and I'm quoting him, he agrees it's a "distortion". It's "wrong". He believes and, in fact, insists that, in the final report, he or the members of the faculty give a true rendition of the findings.'

Then, the growling bear noise of Vasco himself rattled through my earphones: 'The report that I saw contained a quote that came from his presentation,' he said. 'If it's wrong it ought to be corrected. If it's right, it ought not to be corrected.'

And then a staff member hurried to Vasco's defence: 'I find this somewhat disconcerting because I had made copies of Neil's presentation at Millbrae on September 8th, I think it was, and the tone of his voice, the words and everything were very, very positive and very much lively and enthusiastic.'

But as I carried on my research, listening to tapes and reading task force documents, the mystery began deepening again. Dr Smelser was emerging as a rather mercurial character. He was enduringly precise on the detail of the science, and firm in his message that the data wasn't supporting their idea of self-esteem as a social vaccine. But, at times, he also seemed oddly keen to leave an impression that he *agreed* with the task force. He'd make reference, for instance, to 'what we really all know' about self-esteem being a 'key factor' in the genesis of social problems. Then, in a private memo to Vasco, I learned that four days after David's intervention at the meeting, Smelser was contacted by the task force's office. They wanted to see if what David had said at the meeting was true. Was he really angry about the quote they'd used in their public briefing? Did he believe it was a 'distortion' of his beliefs that had to be corrected? Apparently not. Whilst acknowledging it didn't cover all the technical details, the memo recorded Smelser saying, 'I am willing for the quote to stand as it is.' What *was* going on?

I wouldn't discover the truth until I tracked Smelser down.

This was to be yet another revelation. It seemed that, all along, he'd been forced into playing a highly delicate political game. 'My main motive was loyalty to the University of California,' he told me. 'I believed I was bailing them out of a potentially difficult political situation.' It all began when Vasco had called the university's president and asked what they could do to help. 'This was a seemingly innocent phone call,' Smelser said, 'but it was not. Vasconcellos was head of the Ways and Means Committee. The president couldn't ignore this. But he didn't know what to do about it.'

'He couldn't just say no?' I asked. 'He couldn't refuse Vasco?'

'Vasconcellos had power over the university's budget,' he said. 'The pressure was indirect. He didn't say, "I'm going to cut your budget if you don't do it." But, "Wouldn't it be a good idea if the university could devote some of its resources to this problem?"'

The task force, of course, had always treated the involvement of the university as if it was an endorsement of their beliefs. In fact, the opposite was true. Vasco's request had sent a chain of alarm through the university's hierarchy that ended up reaching Smelser. 'They said, "What can we do?" My response was, "Probably nothing." But I thought about it for a couple of days and said, "Well, the university should do what it does best and get involved in a scholarly way."' Vasco, he said, was 'extremely excited' about this plan. 'He thought, "Any attention I could get to this movement is all to the better."' Smelser says he was careful to maintain 'good diplomatic relations' with Vasco and task force chairman Andrew Mecca, even going so far as to name them co-authors of the book. 'So I did all the work and they got their names on the title page.' He described Vasco as 'a weird character. He was impulsive in his behaviour, a very opinionated man, and he sort of frightened you. He was big, a physically large person and had a temper. You sort of had the idea you didn't want to get too close to this guy. But somehow or other he and I developed a pretty good working relationship.'

In the end, the University of California's mission to mollify Vasco set them back $50,000. And Smelser was 'not a bit surprised' when its data didn't support Vasco's beliefs. Of the crucial September 8th meeting, he didn't remember much. 'I was bringing bad news,' he said. 'I'm sure they were disappointed but this was a very civil group.' Their dubious treatment of his data wasn't much of a surprise either. 'The task force would welcome all kinds of good news and either ignore or deny bad news,' he said. 'I found this was a quasi-religious movement and that's the sort of thing that happens in those dynamics.' When I asked about the controversy over the sexed-up quote, and David's contacting him about it, his memory apparently failed. 'I can't remember it,' he said. 'But maybe that's because I don't want to remember it.' And of the delicate balance he'd had to manage between loyalty to the university and the intimidating thunder of Vasco, he feels vindicated by history. 'I think my work was successful, in that regard, because the university was never criticized.'

Next, I decided to contact Bob Ball. He told me, 'It's been over twenty-five years since the task force report was issued . . . at the time I saw no problem with the conclusions and suggestions included in it'. Finally, and for what it was worth, I thought I might as well attempt to trace Andrew Mecca. I was aware of how unlikely it was that the task force's chairman, and former 'Drug Czar' of California, would be candid with me. Not only was he a veteran politician, he'd been Vasco's right-hand man. He wasn't going to admit deliberately misleading the media and public. But why not at least see what else he had to say? When I eventually got hold of him, he confirmed it was the prestige of the University of California that turned things around for the ailing task force. 'That earned us some credibility stripes,' he said. 'All of a sudden it switched from a hokey John Vasconcellos super-liberal touchy-feely undertaking to a scholarly investigation. So that was great.' I was also surprised when he happily admitted what Smelser told me, that the university only became involved in the first place out of fear of Vasco. 'Yep, that's very accurate. John chaired their lifeblood. Their budget!' he laughed.

Then he said something about the scientists' collected work which really startled me. 'As you read the book,' he said, 'it's a bunch of scholarly gobbledegook.'

'What was your response when the data came in and didn't say what you wanted?'

'I didn't care,' he said. 'I thought it was beyond science. It was a leap of faith. And I think only a blind idiot wouldn't believe that self-esteem isn't central to one's character and health and vitality. These behavioural scientists, like Neil Smelser and whatever that did that stuff, it was kind of a last hurrah for some archaic scholars who didn't have a new language and didn't have the courage to dress it up enough for this new consuming market of people out there craving for some guidance around this.' The scientists, he implied, had cynically viewed the entire task as an opportunity. 'On the whole it was kind of like, "Oh we're going to get our old shit published again in a new book."'

'And was Vasconcellos angry when he read their reports?' I asked. 'There was that negative line in Smelser's introduction.'

'Yeah,' he said. 'Well, the thing is, John was an incredible politician and so tenacious. He was pragmatic enough that he felt he had what he needed, and that was a scholarly report that pretty much said self-esteem's important. At least, that's the spin we got in the media. And that was so important because the media were the ones that legitimized it. All of the final stories were quite positive.'

'It was a remarkable thing,' I said.

'Yep.'

It *was* remarkable, especially as David Shannahoff-Khalsa, infuriated by the cover-up, had begun agitating to get Neil

Smelser's deadly quote into the media. He'd included it as part of his personal statement in the report and even faxed it to the cartoonist Garry Trudeau, who'd prepared a new set of Boopsie cartoons. He also managed to get himself quoted in some stories, albeit usually downstream far away from the delirious headlines. 'I think most of it was ignored,' David told me, of his guerrilla actions. 'Maybe there were a few brief mentions. I was probably interviewed by the press five to ten times during all of that.'

I found it truly amazing, the extent to which the media had begun singing Vasco and Andrew Mecca's song. Part of this, Mecca told me, was due to the fact they'd visited powerful editors and television producers up and down the country prior to publication, in a deliberate attempt to construct the story before it could be subverted by David or Neil. 'We made the effort to go around to the editorial boards,' he said. 'John and I could spin a pretty good story together. They knew John, they knew his power, and I could add some considerable enthusiasm. And, remember, then somebody walks right out of that room and crafts the editorial for the next day's paper. That's good spinning.'

An extraordinary $30,000 was spent on their PR campaign. At its height, five publicists were working full time, arranging meetings for Vasco and Mecca and sending out their shrewdly crafted materials. 'What we decided – and we consciously decided to do it – was to make sure that we got out there to tell our story and not let them interpret it from the stuff that was being written by Smelser,' he said. 'We cultivated the message. And it was a very positive one. And that positiveness prevailed.'

'And the object,' I said, 'was to try and deflect the negative voices of Neil Smelser and David Shannahoff?'

'Correct,' he said. 'And most of the headlines in all of the final year, and after the report and then again when the book came out, were basically "Self-Esteem Task Force earns its stripes with scholarly support that says self-esteem is central".'

'And nobody made much of what David Shannahoff was doing?'

'I'm not sure anybody cared,' he said. 'Remember, he wasn't going around telling the story. And one detractor, big deal! He did not get to dominate the story.' Neither did Neil and David's actions much ruffle Vasco. 'There were speedbumps,' acknowledged Mecca, 'but John knew it's how you navigate those speedbumps that defines your self. A couple of remarks were not going to detract from the larger ship as it crosses the sea. And, come on, Will! Who remembers Neil Smelser or Shannahoff? Who? Who remembers them? Nobody! They were tiny ripples in a big tsunami of positive change.'

*

In 2014, a heart-warming letter sent to eleven- and twelve-year-old pupils at Barrowford Primary School in Lancashire went viral on the internet. Handed out with their Key Stage 2 exam results, it reassured them, 'These tests do not always assess all of what it is that make each of you special and unique . . . They do not know that your friends count on you to be there for them or that your laughter can brighten the dreariest day. They do not know that you write poetry or songs, play or participate in sports, wonder about the future, or that sometimes you take care of your little brother or sister after school.'

The international press savoured this deliciously cosy story. It was a life-affirming message from a school that made a priority of nursing the youngsters' self-esteem. Barrowford's teachers were discouraged from issuing punishments, from 'defining' a child as 'naughty' and from raising their voices at them. When dealing with a badly behaved pupil, they were directed to inform them their behaviour was 'mistaken' but to remind them they were 'wonderful'. The school's guiding philosophy, said head teacher Rachel Tomlinson, was that the children were to be treated with 'unconditional positive regard'

– a phrase straight from the mouth of founding Humanistic Psychologist and 'second father' to Vasco, Carl Rogers.

A little more than a year later, Barrowford found itself in the news again. The government inspectorate, Ofsted, had visited. They'd given the school one of their lowest possible ratings. Complaining of 'serious weaknesses', they'd found the quality of the teaching and its exam results to be inadequate. 'Staff expectations of what pupils can achieve are not high enough,' said the report. 'Some staff do not give enough attention to teaching the basic skills of reading, writing and mathematics.' The school, it said, 'emphasised developing pupils' emotional and social well-being more than the attainment of high standards'. Defending her ideas, the head teacher told a local newspaper, 'When we introduced the policy it was after an awful lot of research and deliberation and I think that it has been a success.'

*

In the town of Euclid, Cleveland, on the banks of Lake Erie, it was expected that householders would display the Stars and Stripes, every year, on the 4th of July and Memorial Day. And sure enough, right on cue, in his plain-coloured button-down shirt, khaki trousers and a military-style buzz cut, Rudy Baumeister would pace out onto his clipped suburban lawn and dutifully erect his flag. It was the late 1950s and Rudy was a lifetime employee of Standard Oil. He'd emigrated from Germany after serving with Hitler's army on the eastern front and had spent several months in a Soviet prisoner-of-war camp. Rudy had a high opinion of himself. He was quick to anger and would sometimes punish his two children physically. 'He was very right wing,' said his daughter, Susan. 'He was a control freak. It was always his way, always needing to be the leader. He had a very big ego. From an early age, he was raised to be looked up to. He was the first-born, he was the son, he was the one. He always had to be the one.'

Hiding from all this, in an upstairs room at the back of the house, was a little boy named Roy. Born in the spring of 1953, Roy was fair-haired, blue-eyed, well-mannered and often frightened of his father. His windows looked out over the branches of beautiful oaks and the shelves on his light-blue walls held the books that were his escape. He'd read stories of King Arthur and his noble knights and spend hours poring over his set of *Collier's Encyclopaedia*. At high school, he was top of the class, a valedictorian. But he carried everywhere this painful feeling that he was missing out on all the fun. 'My parents didn't believe that kids should do sports, or go to dances or parties,' he said. 'It was pretty much just come home and do the chores and schoolwork.' The pressure to succeed was vast. 'We had to be the best in everything,' said Susan. 'The tallest, the blondest, the smartest, the most good-looking.' Because Roy found school easy, his parents had him skip the fifth grade. 'It's difficult for a boy to be a year younger than everyone else,' said Roy. 'I grew up feeling like a small person.'

A small person, perhaps, but one with big questions about the world. Somehow, Roy managed to persuade his parents to let him off Sunday school in order that he could pursue his own course of religious instruction. 'It took me three years of Sundays to read the entire Bible. I felt it was really valuable to know what was in it, because it was this mysterious book written by God.' But he was disappointed with what he found. The nearest thing to a satisfying theological discussion was the Letters of St Paul. 'I wasn't persuaded by his argument.' As the 1950s became the 1960s the big questions inside Roy began to press against the walls of his smart colonial-style house. 'As a child, you believe what your parents tell you. But I started realizing some of the stuff they taught me wasn't right.'

In his second year at university he was offered a foreign-study programme, where he took up philosophy. 'I wanted to understand what people are like, why we're here and what we're doing.' He read Freud's *Totem and Taboo*. 'It was a revelation,' he said. 'Freud was approaching the question of where our ideas of right and wrong come from scientifically.' Here, it seemed, was a method by which truths about the secret rules of the human self could be revealed that were superior to religion or philosophy. But when Roy asked his father if he could transfer to psychology, he dismissed the idea. 'You'll be wasting your brain,' he said. He only changed his mind when he discovered that Standard Oil not only employed psychologists but that their salaries were higher than his.

For his sister Susan, Roy's interest in the riddles of human behaviour had an obvious cause. 'Psychology was his way of trying to understand how Dad behaved,' she said. As his education continued, at Princeton, Duke and Berkeley, they'd regularly discuss their father. 'With Roy not being a gregarious person, and being so smart, his way to cope is to analyse,' she said. 'We would try and talk through what he'd been learning, "Well, maybe he's doing this because he's feeling this."' Roy's university years were the 1970s, the era of encounter, rebellion and the Godlike me. 'Understanding and exploring yourself was a big theme in the zeitgeist,' he said. 'It was about getting in touch with yourself. The idea was that, in the past, the establishment had required a lot of conformity and people were repressed. But now you should explore the potential inside you. I was doing a lot of those things.'

Roy's preoccupation with the self-esteem problem at home that he called 'Dad' eventually led him to Nathaniel Branden's *The Psychology of Self-Esteem*. 'I remember being disappointed. He would tell a lot of stories, it was fun to read, but I was looking for science.' But Roy was, nevertheless, to become a true believer. In 1974, he wrote his undergraduate thesis on the different ways people react to public challenges to their self-esteem. At the time, he viewed self-esteem as a 'holy grail: a psychological trait that would soothe most individuals' and society's woes.'

Ten years later, Roy found himself preparing to drive across America with a psychologist named Dianne Tice, a thoughtful and disciplined young woman with a quick wit and long red hair. She'd been awarded a grant to carry out research at Berkeley; Roy had a summer programme at Stanford. He'd suggested they share a ride. It was the year of *Ghostbusters* and the Ray Parker Jr theme song was seemingly never off the radio of his Honda Accord. Roy was thirty-one, tall and handsome and fresh from the implosion of his first marriage. Dianne, too, was bruised by a recent break-up. It was a 2,500-mile trip across a continent. Eighteen hours on the road for five days straight. 'You can't go five days with someone in a small car without either hating them or falling in love,' said Dianne.

In San Francisco, they lived across the bay from each other and would go on weekend trips. 'We went to Big Sur and I almost got swept into a waterfall,' said Dianne. 'A rogue wave knocked me off my feet. There was a bad rip tide that would've sucked me down very far. But Roy grabbed me. He saved me. I said right then, "I owe you my life."' They became partners in love and science. 'We did a lot of that early self-esteem work together,' she said. Although he was not testing its purported benefits, he pursued his work with 'a background assumption' that more was better. Over the course of two decades, Roy probably published more papers on self-esteem than anyone else in the United States.

As Dianne grew closer to Roy, she learned that his difficult childhood in that house on Lake Erie had given the quiet academic an appetite for rebellion. 'He was not allowed to challenge his parents at all, so he built up a reservoir of wanting to do that,' she said. 'They always encouraged him very strongly to challenge the authority of popular culture. "If everybody's doing it, it's probably wrong." So he's always been one to challenge the status quo.' As self-esteem flowered all over the culture, Roy found himself increasingly bothered by this insurrectionist voice. 'When California had their task force to raise self-esteem, I started noticing they were making awfully extravagant claims,' he said. 'Like they could balance the state's budget because people with high self-esteem earn more money and pay taxes.' One day, he picked up a big red book with the names Smelser, Mecca and Vasconcellos written up its spine. What he found in its pages surprised him. 'The data was quite weak,' he said. 'I thought, "If that's the best case, then it's really not that strong."'

In the early 1990s, when the self-esteem movement was becoming a frenzy, Roy was researching a book on the psychology of evil. 'Everybody was saying that low self-esteem was a big cause of violence because people with low self-esteem were aggressive,' he said. 'But I knew from my lab work that they're actually shy and unsure of themselves. They don't want to take chances or stand out. None of that sounds like they're going to be aggressive.' Roy decided to chase down the source of the extremely common low-self-esteem–bully claim. 'Everybody who said it cited somebody else, so I'd look up the previous source, and they'd also cited somebody else,' he said. 'That's when I realized there was no evidence for it. No systematic study had been done.' This was a surprise. 'It would be easy to do that experiment. The fact that it wasn't there made me suspicious.' He began to hypothesize: 'Maybe it's not thinking badly of yourself that causes aggression, but when other people think badly of you. Maybe that's where it all goes horribly wrong.' In 1996, Roy co-authored a review of the literature that suggested it was, in fact, 'threatened egotism' that led to aggression. 'That was a big turning point,' he said. The paper proposed that people who lash out have 'highly favourable views of self that are disputed by some person or circumstance'.

It was an astonishing theory. By now, the self-esteem doctrine had saturated academia as well as the popular culture. Roy's new paper ran counter to everything that society and the experts who inform it were saying. It wasn't low self-esteem that caused violence, he was saying. It was high.

When Nathaniel Branden found out about his study, he was furious. He published an indignant response which complained of 'many instances of specious reasoning', presenting the paper as an example of 'what can happen when consciousness and reality are omitted from the investigation'. For Branden, the violent people Roy wrote of might have appeared confident, but underneath all their bluster, they actually had low self-esteem. 'One does not need to be a trained psychologist to know that some people with low self-esteem strive to compensate for their deficit by boasting, arrogance, and conceited behaviour,' he wrote.

But Roy was only just starting. In 1999 he was invited to lead a team that would review the self-esteem literature in its entirety to see, finally, what effect it had on a host of real-world behaviours, such as happiness, health and interpersonal success. 'I said, "Yes, let's just start."' He was on a sabbatical, at the time, at the Center for Advanced Study in the Behavioral Sciences Unit at Stanford. 'Our first computer search looking for "self-esteem" in the abstract came up with fifteen thousand papers. We had a stack of manuscripts waist-high. Several big boxes full. We cut it down with strict criteria. We wanted actual data, not just clinical case studies and things like that. We sorted through them, critiqued them and tried to pull the information together. We wanted to make sense of it. In what way is self-esteem better or not?'

They discovered a major problem with many of the papers: they relied on self-reporting. 'People with high self-esteem just say everything about them is great. If you give them a questionnaire and ask them about their relationships, they'll say, "Oh yeah, my relationships are great!"' The team decided to only accept papers that measured self-esteem objectively. After their cull of the woolly and the anecdotal, around two hundred remained. Among the most egregious errors they discovered were those in the schooling papers. A correlation had been repeatedly found between high self-esteem and good grades. So, the logic went, if you boosted self-esteem you'd boost grades. But the authors had made exactly the mistake Neil Smelser warned the task force about in his 1988 presentation. 'When they tracked people over time,' said Roy, 'the grades came first and then the self-esteem. It was a result, not a cause.'

Roy began to realize that efforts to boost self-esteem hadn't improved school performance at all. If anything, they'd been counterproductive. Neither did self-esteem help in the successful performance of various tasks. It didn't make people more likeable, in the long term, or increase the quality or duration of their relationships. It didn't prevent children smoking, taking drugs or engaging in 'early sex'. His report made the claims of John Vasconcellos look like those of a street-corner wizard. They did, however, find a few benefits. 'High self-esteem makes you feel good,' said Roy. 'It also seems to support initiative. People with high self-esteem take action because they think they know what's right.' Feel-good hits and initiative. It wasn't much. And as the paper archly observed in its conclusion, 'Hitler had very high self-esteem and plenty of initiative, too, but those were hardly guarantees of ethical behavior.' Roy's study was published in May 2003. 'It was,' he said, 'a shock to a lot of people.'

Following on from his work, today's social psychologists have a much more tempered view of self-esteem. It's now thought that people with too much of it fail more, because they're in denial of their own weaknesses or incompetences. They're also likely to give up challenging tasks sooner, as struggling at anything painfully contradicts their concept of who they are. They've also been found to engage in self-defeating behaviour, prior to a challenge, perhaps so they have a ready excuse when they fail to excel.

As for Roy, rather than fall victim to the traps of self-esteem as his father had done, he would, instead, seek to redefine it. In a paper co-authored with Professor Mark Leary of Duke University, he sketched out what they called the 'sociometer' theory. Self-esteem, went the idea, was a system that monitors how well we're doing in our quest for social acceptance. It tries to keep track of what other people think of us. Assaults on our sense of self-esteem, then, are a form of pain signal that alerts us to the fact that damage is occurring to our tribal reputation. 'Self-esteem is one's subjective appraisal of how one is faring with regard to being a valuable, viable and sought after member of the groups and relationships to which one belongs and aspires to belong,' they wrote.

But the paper also contained a warning. It compared the pleasure of hollow self-esteem boosting to cocaine abuse. 'Drugs take advantage of natural pleasure mechanisms in the human body that exist to register the accomplishment of desirable goals,' they wrote. 'A drug such as cocaine may create a euphoric feeling without one's having to actually experience events that normally bring pleasure, fooling the nervous system into responding as if circumstances were good. In the same way, cognitively inflating one's self-image is a way of fooling the natural sociometer mechanism into thinking one is a valued relational partner.' We have a word for people who have become drunk on their own hollow self-esteem boosting. It is narcissist.

The unique dangers of narcissism first became apparent to Roy in a series of experiments that sought to test his theory that it's people with high self-esteem, not low, who are largely responsible for violent acts. The experiment, which he carried out with the psychologist Professor Brad Bushman, involved two people playing a game in which the loser is punished with highly unpleasant barrages of noise. Each player sets the level at which their opponent gets blasted. Would those with high self-esteem, as Roy predicted, turn the sound up to the most aggressive levels? Actually, no. When they checked the data they found the effects of self-esteem on aggression to be surprisingly weak. It was confusing. But the participants had also been measured against a different personality test. 'People had just started talking about narcissism, which seems to be the nasty kind of high self-esteem,' he said. 'That had the strong effects. It was people who were high in narcissism who were more provoked and more aggressive than everybody else.'

The study illuminated the fact that high self-esteem is a mixed category. Some of their participants were presumably healthily and accurately confident in themselves. Their high self-esteem was justified. Their sociometers were functioning well. They were great people, and therefore unlikely to be naturally violent. 'If you went up to Einstein and told him he was stupid,' said Roy, 'he's not going to get mad.' Likewise, if Jesus told you he was a popular guy you could hardly argue he was mistaken. But narcissism is different. 'It's the desire to feel you're superior,' said Roy. 'Narcissists believe they deserve to be treated better than other people.'

It is, Roy has written, 'possible to see narcissism as a kind of addiction to self-esteem'. So what would happen if you took an entire generation of young people and systematically and repeatedly masturbated their self-esteem mechanisms by telling them they were wonderful and special? Could the children of Rogers, Rand, Branden and Vasconcellos be growing into a generation of narcissists?

*

It was 1999. In a basement office in the lab of Roy Baumeister two ambitious young psychologists that he'd mentored were killing time in the boring Cleveland winter. Jean Twenge talked about some of the ways changes in US culture seemed to be making themselves apparent in personality tests, whilst W. Keith Campbell mulled over his research into narcissistic behaviour. Suddenly, the idea seemed obvious. Why not put their interests together? Why not see if narcissism in America had changed along with the culture?

It would be a while until the all necessary data became available. But eventually, in 2008, they published the details of what they'd discovered in the *Journal of Personality*. They'd completed a meta-analysis of eighty-five studies that included the narcissism data from an impressive 16,475 college students, stretching back to the early 1980s. These students had taken the Narcissism Personality Index, or 'NPI', a test used widely by psychologists to measure narcissistic traits. Their paper, 'Egos Inflating Over Time', suggested that, between 1982 and 1989, average NPI scores actually went down, from 15.55 out of a possible 40 to 14.99. But in 1990, when Vasco's final report was published, they started to wobble.

That year, one study pegged them at a relatively low 14.65 but another result marked a new high: 15.93. Four years later, one study reached another new high: 17.89. Then 1999 saw yet another record broken: 19.37. By 2006, one study had come out at an extraordinary 21.54. When the data from all eighty-five papers were averaged out, NPI scores over that period had risen by around two points. This, according to the authors, was a significant jump in a relatively short space of time. It meant that 'almost two-thirds of recent college students are above the mean 1979–1985 narcissism score, a thirty per cent increase'. By the mid-2000s, when the children of the self-esteem generation had become parents, the problem was accelerating. Narcissism, the researchers argued, was now an 'epidemic', rising as quickly as obesity. The increases they'd discovered were equivalent to 'the height of all men going up by about an inch'. The irony was intense. 'Narcissism causes almost all of the things that Americans hoped high self-esteem would prevent,' wrote Twenge and Campbell, 'including aggression, materialism, lack of caring for others, and shallow values. In trying to build a society that celebrates high self-esteem, self-expression, and "loving yourself," Americans have inadvertently created more narcissists.'

But Twenge and Campbell's work soon came under very public attack. Some critics claimed that people in every generation exhibit narcissistic behaviour until they grow out of it, and that these judgements were just the usual grumbles of the middle-aged directed at the young. This would've been an elementary error if Twenge and Campbell had actually made it. But, of course, they hadn't. For a start, said Twenge, 'there does not exist a longitudinal study that has followed people from young adulthood, to middle adulthood to see what happens to their narcissism scores. It doesn't exist. There's certainly other data that points in the direction that younger people are probably going to be more narcissistic, but that's the whole point of using the method of over-time data. We're not comparing eighteen-year-olds to fifty-year-olds, because if you do that you can't tell if it's age or generation, so you really need samples of eighteen-year-olds going back in time or middle school students going back in time. And that's what we've been doing since the beginning, so it's very odd that anyone would provide that as an alternative explanation because it's so obviously untrue.'

As for the simpler argument that older people always view the younger as more narcissistic, 'I hear that argument all the time,' she said. 'It's based on the premise that all findings of cultural change and generational change are based on the perceptions of older people, but virtually none of it is. Almost everything I've published has been about what young people say about themselves.'

A more serious attack came from Dr Kali Trzesniewski of the University of California, Davis. She was sceptical that a two-point rise in NPI scores could really be described as an 'epidemic'. She also claimed to have re-analysed the data Twenge used with greater precision and found no rise in narcissism. As well as this, she cited a new dataset, the Monitoring the Future Project, which assessed high-school seniors every year between 1976 and 2006. That found no rise in self-esteem. Finally, she presented her own analysis of narcissism in Californian universities which, again, found no rise.

But Twenge pushed back. The assertion that the rise in narcissism flattened when the data was treated with greater precision was simply 'untrue', she told me. When Trzesniewski's own data on Californian universities was published, Twenge analysed it herself. She was surprised to find it compared the narcissism levels of students at different institutions at different times. 'All of their early samples are from Berkeley. All of their later samples are from UC Davis. Davis students happen to score significantly lower in narcissism than Berkeley students, so it's like you're comparing the heights of men in one year with the heights of women in a later year and concluding there's been no change in height.' When she re-analysed Trzesniewski's data, by charting the increases in narcissism at the University of California, Davis, alone, 'My jaw hit the floor,' she said. 'Between 2002 and 2007, the NPI scores at Davis went up every single year in lock-step. We usually don't see that sort of change over that short a period of time.' Trzesniewski continues to dispute Twenge's work, and stands by her findings. 'I obviously feel the data I've presented is stronger or I wouldn't be publishing it,' she told me.

As for her final attack, Twenge said it had more merit. The Monitoring the Future Project did indeed find no rise in self-esteem. 'But that's science,' said Twenge. 'It's messy.' There are, she countered, now 'eleven studies showing a generational increase in narcissism, seven using non-college samples. They include respondents from high-school age to adults, four different ways of measuring narcissism, three different research methods, four different ways of recruiting respondents, three different countries and eight sets of authors. This is an overwhelming amount of evidence. Nor is the increase subtle. A 2008 study found a tripling in prevalence of Narcissistic Personality Disorder. Fifty-eight per cent more college students answered the majority of NPI items in the narcissistic direction in 2009 compared to 1982. Thus, twenty-two studies or samples show a generational increase in positive self-views, including narcissism, and only two do not.'

If all that wasn't enough, what we already know about the

form and development of the self surely predicts some sort of effect from the cultural changes that had been taking place. If we are what we think other people think we are, and other people keep telling us they think we're unique, talented and special winners, then that's what many of us are going to believe, at least to some extent. Given that these ideas are aimed mostly at the young, in whom the sense of self is both vulnerable and formative, it would perhaps be more surprising if there hadn't been the kinds of upticks that Twenge and Campbell had reported.

Evidence that parental overpraise is, indeed, a cause of narcissism was presented in an impressive 2015 study in the Proceedings of the National Academy of Sciences. Its aim was to test two long-competing theories. The first was the Nathaniel Branden-esque idea that children become narcissistic when their parents are cold and rejecting. The other theory insisted that, on the contrary, children become more narcissistic when parents over-value them, telling them they're special and better than everyone else. The team, led by Dr Eddie Brummelman at the University of Amsterdam, assessed the levels of narcissism in 565 children as well as the parenting styles of their caregivers every six months over a two-year period. 'We found very strong support for the second idea and not for the first,' he told me. 'The more parents overvalued their kids, the more narcissistic their kids had become six months later. There was no evidence that lack of parental warmth or affection predicted narcissism.'

But over-praise isn't the only problem. Following the self-esteem revolution, parents and teachers increasingly sought to protect children's esteem by artificially cocooning them from natural consequences. One manifestation of this is the grade inflation that's been reported in educational institutions in America and Britain. Between 1968 and 2004, SAT scores for college applicants in US schools fell – and yet the proportion of college freshmen claiming an A average in high school still somehow managed to rise from 18 per cent to 48 per cent. In 1999, in the UK, 8 per cent of graduates received a first-class degree. By 2017, that proportion had increased to 25 per cent, a substantial and rapid rise that's hard to pin solely on students performing better. In 2012, the chief executive of the British official exams regulator Ofqual admitted the value of GCSE and A-Levels had been eroded by years of 'persistent grade inflation' that was 'impossible to justify'. None of this, said Campbell, is very useful for young people. 'Burning yourself on a stove is really useful in telling you where you stand,' he says, 'but we live in a world of trophies for everyone. Fourteenth place ribbon. I'm not making this stuff up. My daughter got one.'

How much of all this would've happened anyway, if John Vasconcellos hadn't told his lie? After all, as we've been tracking from 500 BC, the economy and society of the West were certainly moving, with ever-increasing mania, in the direction of the great, glorious I. One study, also by Twenge and Campbell, found self-esteem levels (which isn't narcissism, remember) in students rising as early as the 1960s, with rates in children and early adolescents picking up from 1980, following a drop in the 1970s. The wider application of school programmes can be traced back to later in that decade, presumably at least partly inspired by the success of Nathaniel Branden's book. It's not possible to replay history minus this weird element, so we can never know for sure. But the story told by the contemporary news accounts is, I believe, strongly suggestive that Vasco's lie had an impact upon the culture of America that inevitably soaked into Britain and elsewhere. It appears that self-esteem, prior to the professors' 'endorsement' of Vasco's task force, had been reasonably well known but seen as rather niche. It was also, in its Rand–Branden phase, focused much more around hard work and achievement, rather than simple belief in individual specialness. And then the conversation changed.

One final dip into the news reports of the time is revealing.

In the early years of the 1990s self-esteem boom, journalists naturally started asking the question, why is all this happening? A 1990 story on a suddenly popular 'Self-Esteem Fair' said its organizer gave 'much of the credit for the rapid rise of the movement to Trudeau's lampoon in his Doonesbury strip of California's Task Force for Self Esteem. "If it weren't for Trudeau's two week stint," he said, "we figure we wouldn't have received even half of the calls we have received in the last year and a half."' A lengthy *New Woman* article in 1991, called 'SELF-ESTEEM: THE HOPE OF THE FUTURE', claimed it was 'the highly publicised work of the task force' that had 'fuelled the growth of the burgeoning self-esteem movement'. A year later, a *Newsweek* cover story, largely critical of the movement, said, 'The man most responsible for putting self-esteem on the national agenda is not a clergyman or philosopher but a California assemblyman named John Vasconcellos.'

*

Alan Greenspan had a problem. In January 1993, the arrival of Bill Clinton in the White House presented a threat to his grand project of maximizing the forces of competition. Clinton had campaigned on the promise of a gentler form of neoliberalism. Whilst continuing to honour the freedom of the individual and the markets, he wanted to take better care of the people, especially those who the game of neoliberalism was hurting the most. He'd promised to help them by investing in education, health, and infrastructure . . . and he'd won. Greenspan met the new president, for the first time, shortly after his victory. At a private meeting at the Clinton residence in Little Rock, he tried convincing him that his focus shouldn't be on the generous public investments he'd been promising. If he was going to avoid the kind of economic calamity that was last seen in the 1970s, he must instead do everything in his power to pacify Wall Street. It was imperative that they be convinced that he wasn't going to spend irresponsibly and thus deepen the national deficit. The meeting went on for one hour, then another, and eventually merged into an unplanned lunch. By the time he'd left, Ayn Rand's protegé had Clinton spooked.

Less than two weeks before his inauguration, at the first gathering of his National Economic Council, Greenspan's Vice-Chairman continued the pressure. As the message, that his first duty should be to Wall Street rather than Main Street, was pressed once more, Clinton's face is said to have pinked with rage. He hissed to his people, 'You mean to tell me that the success of my program and my re-election hinges on the Federal Reserve and a bunch of fucking bond traders?'

On the eighth day of his presidency, over an intimate dinner at the Metropolitan Club in Washington, Greenspan met with Clinton again. He warned of a 'financial catastrophe' after 1996 if he didn't comply with his plan. Eight days later he increased the pressure yet further by saying he needed the national deficit cut by an incredible $140bn. On 17 February, in his first State of the Union address, Clinton announced his new economic plan. Warning of a national debt that, if stacked in thousand-dollar bills, 'would reach 267 miles', he insisted that America had to learn to live within its means, telling the people, 'each of us must be an engine of growth and change'. Sitting next to Hillary, in seat A6, was a politely applauding Alan Greenspan.

His power was now immense. Enthralled by the lessons he'd learned at Ayn Rand's knee, where he was taught it was the capitalist's very profit-seeking that was the 'excelled protector of the consumer', he pushed for a dramatic expansion of the neoliberal game by advocating the deregulation of the financial industries. In 1999, Clinton repealed the laws that had been brought in, in 1933, to control the banks following the crash – laws that had helped kick-start the now long-vanished period of Great Compression. This wave of deregulation brought into being the highly unstable derivatives market that was made up, in the words of superstar investor Warren Buffett, of 'financial weapons of mass destruction'. From a starting position of almost nothing, those weapons of mass destruction quickly became a $531tn industry.

Whilst all this was happening, the low interest rates that were another of Greenspan's preoccupations enabled millions of cash-strapped people to take on irresponsible levels of mortgage debt. In 2004 he was hailing the 'resilience' of the financial system; in April 2005, he voiced his approval of the new and thriving 'subprime mortgage market'. 'Where once more-marginal applicants would simply have been denied credit,' he said, 'lenders are now able to quite efficiently judge the risk posed by individual applicants and to price that risk appropriately.'

Freedom! Freedom to imperil the world's banking system with highly unstable products. *Freedom!* Freedom to take out mortgages you can never hope to repay. Ever since the days of Ancient Greece, the West had fetishized freedom. Since then, the default truth of our people has always been that the locus of power is the individual, and that what that individual requires to succeed is maximum freedom, whether that be freedom from the repressive forces of society, from self or from regulation by the state. It was a collision of such freedoms that did much to cause the 2008 global financial crisis. This was a neoliberal catastrophe. As a result of the crash, in the US alone, more than 9 million homes were lost and almost 9 million jobs disappeared. In the UK, 3.7 million people lost their jobs – one in seven of all employees.

A few months after my visit to the Suicide Lab, and my conversation with Rory O'Connor about the mysterious rise in male suicides, answers began finally to appear. Researchers concluded that, in the UK, between 2008 and 2010, there were around a thousand *extra* suicides, nearly 90 per cent of them male, because of the crash and the period of austerity that followed. There were also an astonishing 30–40,000 more suicide attempts. As bad as that is, the true scale of the human disaster can only be glimpsed by looking at the global figures. In the same period, according to a study in the *British Journal of Psychiatry*, the estimated number of additional suicides comes out at 10,000.

Ten thousand people dead. If the violence had been direct, this massacre wouldn't have been so easily ignored. It was, of course, the Chairman of the Federal Reserve, Ayn Rand disciple and devil in Clinton's ear, Alan Greenspan, that was a central character in its bloody story. In the aftermath of the crisis, the *New York Times* had this to say: 'Over the years, Mr Greenspan helped enable an ambitious American experiment in letting market forces run free. Now, the nation is confronting the consequences . . . He counted among his formative influences the novelist Ayn Rand, who portrayed collective power as an evil force set against the enlightened self-interest of individuals. In turn, he showed a resolute faith that those participating in financial markets would act responsibly. An examination of more than two decades of Mr. Greenspan's record on financial regulation and derivatives in particular reveals the degree to which he tethered the health of the nation's economy to that faith.'

Of course, it's important to remind ourselves of the complicated, grey and grown-up truth that neoliberalism isn't an evil villain. It's not moral. It's not trying to be fair. It's a system. Like most systems, it creates a trade-off in outcomes. It has both positive and negative effects. Many in the West have become wealthier since the 1970s and their standards of living have risen. Global free trade has helped lift millions more out of poverty in developing countries such as China, Indonesia and India. Around the world, the number of people living in extreme poverty has declined by more than a half since 1990. Globalization – an integral component of the neoliberal project – has been a blessing for many.

But one of neoliberalism's most negative effects is its tendency to concentrate the pain on our most vulnerable. For the game to properly function, it naturally requires winners and losers. The more the winners win, the more powerful they become, and the more able they are to tweak the game's rules in their favour. Ever greater freedom from regulation, for business and banking, has meant those businesses and banks have been increasingly empowered to act in their own self-interest – and all this has led to an astonishing rise in inequality. Prior to the 1940s, the wealthiest 1 per cent of the US population were receiving 16 per cent of the national income. During the years of the Great Compression, their share fell to less than 8 per cent. Between 1979 and 2009 it rose to 23 per cent. Between 1978 and 2014, inflation-adjusted CEO pay increased by nearly 1,000 per cent. Meanwhile, globalization has brought with it the astonishing rise of the power of multinationals such as Apple, Siemens and Exxon-Mobil. In 2009, economists reported the startling fact that forty-four of the world's hundred largest economies were not countries, but corporations.

Neoliberalism's disdain for regulation and government oversight has meant the rich and powerful have been free to make themselves even more richer and powerful – even during the devastation of the financial crisis. In the post-crash years of 2009 to 2012, median pay for the US worker fell by 3.4 per cent. In the UK, meanwhile, between the crash and 2013, average family income fell by 3.8 per cent. And yet over the course of just one year, 2010–11, CEOs in the US enjoyed a 40 per cent rise. In the UK, directors of the top 500 companies bumped their pay by 49 per cent.

Today, house prices are rocketing, personal debt is soaring and wages are stagnating. Between 1980 and 2014, US consumer credit as a percentage of household income nearly doubled. American students owe more than $1.2tn in college debt, whilst in Britain they're £73.5bn in the red. And, still, state protections are being dismantled. Globalization has led to cheap imports and higher levels of immigration. Industry is embracing harsh new working conditions, such as 'zero hours contracts', which offer minimal job security or benefits. According to the Office of National Statistics, 800,000 people in the UK were on zero hours contracts in December 2015, up by 15 per cent on the previous year. A 2016 study found 4.5 million people, nearly one in six of all workers, in England and Wales, were in 'insecure work.' In the US, meanwhile, the number of neighbourhoods classed as 'extremely poor' leapt by 45 per cent following the crash, whilst the number of people living in them rose by 57 per cent. By 2015, one in five US households lived either in or near poverty – 5.7 million Americans had joined that number since the financial crisis.

Perhaps most tellingly of all, neoliberalism has honed the way corporations view themselves and their role in society. Back in 1981 the American Business Roundtable declared that 'the longterm viability of a corporation depends upon its responsibility to the society of which it is a part', a statement which seems rather quaint just a few decades on. These days, when business leaders are challenged about the lengths they go to reduce their tax bills, for example, they frequently defend their actions by presenting themselves as ultra-smart contestants in the neoliberal game who are managing to reduce their liabilities whilst still cleverly playing within its rules. In 2016, a three-year investigation by the EU found that electronics company Apple Inc. had paid between 0.005 per cent and 1 per cent tax on their European profits. The commission found that their 'sweetheart' deal with Ireland, where they'd based their European HQ, amounted to a form of illegal state aid and attempted to claw back $13bn. Dismissing all this as 'political crap', Chief Executive Tim Cook defended Apple by explaining that they'd 'played by the rules'. Similarly, following the financial crisis, some bankers in Wall Street saw their bailing out by the state as proof of their superior skill at the game. 'The fact that I benefited from [the bailout] is because I'm smart,' one told *This American Life* reporter Adam Davidson. 'I took advantage of a situation. 95 per cent of the population doesn't have that common sense. The only reason I've been doing this for so long is because I must be smarter than the next guy.'

Connected to this 'you are not our responsibility' corporate mindset is our age of perfectionism's distinctive 'no-service' consumer culture, in which we're increasingly stuck in byzantine call trees, thrown towards volunteer online forums for technical support and checking ourselves into flights and our shopping out of supermarkets – the constant, tedious drip-drip-drips of the market, adding more and more stress and responsibility onto the teetering 'I'.

One place this hyper-individualist model of the corporate self is dominant is Silicon Valley. As veteran tech reporter Dan Lyons has observed, employees there are often reminded that the company they serve is not their 'family' but their 'team'. This definition was introduced by Netflix in 2009 and has since spread rapidly throughout the sector. 'The Netflix code inspired a generation of tech start-ups and "may well be the most important document ever to come out of the Valley," Facebook COO Sheryl Sandberg once said,' writes Lyons. 'The result, according to countless articles in publications like *Fortune*, the *New Republic*, *Bloomberg* and *New York Magazine*, is that Silicon Valley has become a place where people live in fear. As soon as someone better or cheaper comes along, your company will get rid of you. If you turn fifty or forty, or thirty-five; if you demand a raise and become too expensive; if a new batch of workers comes out of college and will do your job for less than what you are paid – you're gone. So don't get too comfortable.'

*

As the financial crisis was shaking the world, sales of Ayn Rand's most famous work, *Atlas Shrugged*, began booming; 2009 saw purchases triple over 2008. In 2011, 445,000 copies were sold, surpassing even its bestselling year of publication. In 2012, Wall Street journalist Gary Weiss wrote: 'Rand has experienced an extraordinary revival since the financial crisis, and nothing seems to be stopping her. It's a struggle for the soul of America, and she is winning.'

Five years later, she'd have her greatest victory yet. A man who had, as a role model, the uncompromisingly individualistic hero of her 1943 novel *The Fountainhead* would be sworn in as the 45th President of the United States of America. Donald Trump was a billionaire businessman who'd once praised Rand's book in grandiose terms, saying, 'it relates to business, beauty, life and inner emotions. That book relates to . . . everything.' And the new president endorsed like-minded friends, nominating a secretary of state, a secretary of labor and a director of the CIA who were all avowed fans of Rand and her ideas.

Trump was, in many ways, a definitive creature of the neoliberal, self-esteem, celebrity era. A sumptuously narcissistic self-publicist, he'd initially become famous during the 'Greed is Good' 1980s only for his fame to become supercharged when he starred in the immaculately neoliberal reality show *The Apprentice*, which turned business into a ferociously competitive game in which losers suffered the public humiliation of a boardroom beat-up and his catchphrase, 'You're fired!' But while his television profile was an important component of his rise to the heights of power, any true understanding of it would be incomplete without an account of one of his most effective tools: the echo chamber and sounding board of social media.

Telling that story means driving roughly two and a half hours north-east of the Esalen Institute, to the district that John Vasconcellos served faithfully and doggedly, as Assemblyman, for thirty-eight years. It's in this bland landscape of offices and suburbs that Ayn Rand is something of a secret hero. And it's here – California's Silicon Valley – that will be our final destination in the journey of the Western self.

If the 'age of perfectionism' is one filled with opportunities for us to feel like failures, then the rise of neoliberalism, the financial crisis and the harshly competitive world it's all left us with is clearly a major part of it. But another is the internet and social media, which have become increasingly dominant, in our time, especially in the lives of the young. It's in this milieu that the self-esteem generation has become the selfie generation.

BOOK SIX

The Digital Self

Black screen. Polite applause. A man, Doug Engelbart, appearing in a headset. He wore a black tie, fastened high and tight; his eyes in shadow beneath heavy brows that gave him an air of distance and melancholy. Many of the technology experts in the audience considered Engelbart a 'crackpot'. They had no idea he had, upon his desk that day, pieces of technology that were as if from a time machine. Engelbart was touching the future, and he was about to show them it.

'I hope you'll go along with this rather unusual setting and the fact that I remain seated when I get introduced,' he said, up on the screen. 'I should tell you I'm backed up by quite a staff of people between here and Menlo Park where Stanford Research is located, some thirty miles south of here and, er,' he smiled anxiously and glanced upwards at some unseen person or thing, 'if every one of us does our job well, it'll all go very interesting.' He looked up again. 'I think.' Another nervous pause. 'The research programme that I'm going to describe to you is quickly characterizable by saying if, in your office, you as an intellectual worker were supplied with a computer display backed up by a computer that was alive for you all day and was instantly responsible.' A glance up. Embarrassment. 'Responsive. Ha. Instantly *responsive* to every action you had, how much value could you derive from that?'

What the man on the screen was describing, at the Fall Joint Computer Conference in San Francisco on the morning of Monday 9 December 1968, was a personal computer. To most of the people present this was an utterly novel concept. In the years of the Great Compression, with its collective economy that was so deviant to the American subconscious, a general neurosis had arisen amongst the population that said that computers weren't there to serve the individual, but the organism. At that time, Silicon Valley was the brain of the military–industrial complex, a place of secretive laboratories working on defence contracts: missiles, radar, guidance, nuclear weaponry. Computers were great humming mainframes, mysterious, alien, basement presences that were operated by teams of inscrutable people in lab coats. They were machines of conformity and control and were to be feared. Even the experts in the Valley, many clustered around Stanford University, saw the future as one in which computers would replace humans, believing true artificial intelligence was coming soon. But Engelbart's vision was radically different. And so was the technology he was about to demonstrate to the stunned crowd.

The glare from his monitor glowed onto his face as he explained that they'd been developing this new form of computing at the Stanford Research Institute in Menlo Park. 'In my office I have a console like this and there are twelve others that have computers and we try, nowadays, to do our daily work on here.' He smiled as if in acknowledgement of how eccentric all this sounded. 'So this characterizes the way I could just sit and look at a completely blank piece of paper. That's the way I start many projects. I'll sit here and say, "I'd like to load that in . . ." Off screen, he moved his right hand. There was an odd clicking sound. Then he began typing. On the screen, characters – 'WORD' 'WORD' 'WORD' – began magically appearing. 'And if I make some mistakes I can back up a little bit.' Just like that, the words began to disappear. 'So I have some words and I can do some operations with them.' A cursor appeared, a black dot zipping about like a maddened gnat. 'I can copy a word. In fact, there are a pair of words I'd like to copy – and I can just do this a few times.' He copy-and-pasted WORD, WORD, WORD, WORD. 'So I could get myself some material on a blank piece of paper and then I'd say, well, this is going to be more important than it looks, so I'd like to set up a "file". So I tell the machine, "Output to a file." And it says, "I need a name." I'll give it a name. I'll say it's "sample file".' The screen faded to a shot of Engelbart's hands. He was working on a typewriter-like QWERTY keyboard that was connected by wires to the monitor on which the words had been appearing and disappearing. To his right was a strange box containing wheels that he used to move the cursor. He called this contraption a 'mouse'.

When it was all over, the crowd rose to their feet and cheered, spellbound, enthralled. Not only had Engelbart introduced the world to the notion of the computer as a personal assistant controlled by a mouse, keyboard and cursor, he'd shown them a graphical user interface which formed the basis of the 'windows' he'd been manipulating, hyperlinks and the concept of the networked online realm we know today as the Web. It would become known, in Silicon Valley, as 'the mother of all demos'. Up on that screen, Engelbart the crackpot had, in the memorable phrase of one observer, been 'dealing lightning with both hands'. It wasn't just that the technology was new, the foundational concept behind it was revolutionary. Here, for the first time, was computing that had been designed to be personal. It wasn't a machine for repression and coercion. It all worked in service of the I.

Eighteen years earlier Doug had found himself driving to work in a state of crisis. A wartime naval technician, he'd been recruited by the forerunner of NASA to work at Silicon Valley's Ames Research Center. He was twenty-five and content in work and love. What more was there do to? All his goals were met. This insight came to him during his commute and he found it so awful he'd had to pull his car over. 'My God, this is ridiculous. No goals!' He spent that night preoccupied with finding a new project to aim his life at. He knew he wanted to change the world. But how? Could he invent something that would help humans cope better with the dizzying complexity of the future?

It all came to him, then, in a torrent of fabulous insight that Silicon Valley historian John Markoff has called 'a complete vision of the information age'. In his vision, he saw a man sitting at a screen that was attached to a computer and there were characters on the screen and the computer would be a kind of portal to all the information you'd need to do your work, and it would be wired to other computers, so you could communicate with each other, and your computer would work *for* you, organizing and aiding your working life. It would be a 'tool for thought', an extension of the mind. This was, of course, a deeply individualistic idea. But it was also humanistic – the Carl Rogers insight transposed onto the realm of tech. Computers weren't bad, they were *good*. They weren't there to control the masses, they'd *free* them by augmenting the incredible potential that humans already possessed.

'Augment'. It was a good word. After studying electrical engineering at the University of California, Berkeley, Engelbart began working at the Stanford Research Institute (SRI). There, he wrote several papers, including 'Augmenting Human Intellect', which detailed his notion of workers connected to each other by computers which they'd control 'by means of a small keyboard and various other devices'. In 1963, he received funding from the Defense Department to develop his ideas. And so, on the second floor of the SRI in Menlo Park, he established the Augmentation Research Center (ARC). The team he recruited to ARC would not only invent the future of Silicon Valley, they'd even *look* like its future. Writes Markoff, 'In the midst of this engineer's world of crewcuts and white shirts and ties arrived a tiny band distinguished by their long hair and beards, rooms carpeted with oriental rugs, women without bras, jugs of wine and on occasion the wafting of marijuana smoke. Just walking through the halls of the SRI lab gave a visitor a visceral sense of the cultural gulf that existed between the prevailing model of mainframe computing and the gestating vision of personal computing.'

But this cultural mash was to become one of ARC's greatest problems. It's partly responsible for the fact that Engelbart isn't world-famous today. Despite the brilliant reception he won at the 1968 conference, the next few years would see the relationship between his lab and his bosses foundering: why, wondered the Defense Department, were they funding some sort of electronic office assistant? And why, asked the programmers who'd joined his project with hopes of improving the world, were they developing tools for the Defense Department? Further strife came in the early 1970s when a few key members left for the new Palo Alto Research Center (PARC) that had been set up by the photocopying firm Xerox, which was also now interested in personal computing. And then, into this growing discord, Engelbart introduced encounter groups.

By this time, the Esalen Institute, 130 miles to the south, had become an influential force around Stanford. It was the university that had created SRI (and would separate from it in 1970) and from which Esalen's founders had graduated. In 1968, the year of the demo, the Institute's co-founder Michael Murphy had written in the *Stanford Alumni Almanac* of 'several professors', including teachers of electrical engineering, who were 'incorporating new disciplines for self-exploration into their ongoing classes'. Workshops based on Esalen programmes were highly popular at the university, having been attended by eight hundred members of the community. These programmes became so successful that funding was received to teach Esalen techniques across the Bay Area at elementary and secondary schools, with more than fifty thousand participants. Esalen would remain popular with tech workers for many years, with one 1985 *Esquire* story reporting 'scientists, engineers and entrepreneurs from Silicon Valley' being regular guests. Such was its lasting influence, Tim O'Reilly, the man who in 2005 christened the internet's social media phase 'Web 2.0', would eventually comment, 'The internet today is so much an echo of what we were talking about at Esalen in the 70s.'

Engelbart's interest in encounter had come via the Erhard

Seminars Training workshops or 'EST' that Esalen (in particular its gestalt programmes) had partly inspired. Its introduction to ARC began with a three-month experimental period involving thirty staff. The radical authenticity that encounter brought with it unbottled a flume of dissatisfaction, much of which was directed at Engelbart. 'There is an impression that Doug goes off in a corner and hatches ideas. People are uncomfortable with all the surprises . . . Doug does not allow enough control, goal setting, participation in general.' At the experimental period's end, an evaluation was ordered. The final report concluded, 'Frankly we don't know what the hell is going on.'

But Engelbart pressed on regardless, convinced EST and encounter would help his team become more creative. Pressure was put on the staff to attend courses, half of the cost of which was paid by the lab. After the first cohort returned from theirs, bug-eyed and evangelistic, more and more began to succumb. A cult-like fog descended. Authenticity spread through ARC like a foul burp: there was a wave of divorces, with one programmer being told by his wife she'd been sleeping with his best friend. Some programmers became hyper sensitive and difficult to deal with, others moved to communes and others still gave up computer research altogether. Visiting suits from the Defense Department would arrive to find barefooted staff members sitting in a circle, wine bottles open, spliffs lit, mid-encounter. It did little for their working relationship.

Staff members who'd managed to hold out against EST began to find the working environment intolerable. One, Donald 'Smokey' Wallace, complained that Engelbart was treating his people like 'laboratory animals' and he, along with a handful of others, resigned. Inevitably, in 1974, Engelbart's funding was pulled. It would be Xerox, with the help of various Engelbart alumni, that would turn his ideas into reality by improving it, over time, with the benefit of newer technology and an emphasis on simplicity. In 1979, they'd demonstrate their work in a now notorious private demo for an ambitious tech entrepreneur called Steve Jobs – who raced back to his company, Apple Computer, and ordered his people to begin copying it. But it would be a while until the rest of the planet, including Jobs, caught up with Engelbart's genius completely. The mouse wouldn't become available to the wider public until 1983 and his astonishing vision of the internet would take even longer to actualize. After seeing computers networked together at the Xerox demo, Jobs had largely shrugged off this function, so taken had he been with the user interface. Engelbart recalled meeting Jobs in the 1980s. According to his account, he told Jobs his Apple computer was 'terribly limited. It has no access to anyone else's documents, to email, to common repositories of information.'

But Jobs didn't get it. 'All the computing power you need will be on your desktop,' he said.

'But that's like having an exotic office without a telephone or door,' replied Engelbart. Jobs, he said, just ignored him.

Much of the spirit of Engelbart's wider vision would be carried into the future by the man who was not only an invited speaker, along with Vasco, at Esalen's bad-tempered 'Spiritual Tyranny' conference, but who acted as consultant and cameraman for the 1968 demo. To Engelbart's technology, Stewart Brand would add the humanist-neoliberal ideology that still drives our computer culture today. The month following the demo saw the first proper publication of Brand's *Whole Earth Catalog*, whose headquarters was blocks from Engelbart's lab. Open the cover of this esoteric bazaar of products and philosophies and you'd see its mission statement. Declaring 'we are as gods and might as well get good at it', it hailed a future in which 'a realm of intimate, personal power is developing – power of the individual to conduct his own education, find his own inspiration, shape his own environment, and share his adventure with whoever is interested.' Brand's biographer, Stanford University's Professor Fred Turner, would archly note that the *Catalog* presented a way of changing the world through buying things, an 'idea which has stuck around'. In 2005 Steve Jobs called the *Whole Earth Catalog* 'one of the bibles of my generation . . . sort of like Google in paperback form'. The instruction on the final page of its final issue, 'Stay Hungry. Stay Foolish', would be a mantra that guided his life and career.

Brand and his cohorts are often depicted as kind of techno-hippies. But what's not often noted is that their utopian vision of a world without hierarchy, in which the old, centralized, authoritarian orders would be washed away, leaving all the individuals free to manage their lives as they wished, was strikingly similar to Friedrich Hayek's own vision of neoliberalism, in which the old centralized, coercive state would be crushed to its minimum, leaving all the individuals free to manage their lives as they wished. In 1995, Brand told *Time Magazine* it was the counterculture's 'scorn for centralized authority' that provided the philosophical foundations of 'not only the leaderless internet but also the entire personal-computer revolution'. Indeed, his Greek heart had been in evidence since childhood. As a student, terrified of the Communists, he'd written in his journal, 'if there's a fight then I will fight. And I will fight with a purpose. I will not fight for America, nor for home, nor for President Eisenhower, nor for capitalism, nor even for democracy. I will fight for personal individualism and personal liberty.' He lionized the entrepreneur, the individualist self-starter, and had an authentic neoliberal's distaste for altruistic government, later writing, 'That whole victim mindset – saying to the government you're supposed to fix my problem – is a total anathema to Whole Earth.'

But as much as Brand himself, it was the network of like-minded influencers he drew around him that would have a defining effect on the digital culture of today. From the *Whole Earth Catalog* sprouted a range of other magazine titles and a new form of community that gathered 'online', connected by computers and modems, that he called Whole Earth 'Lectronic Link. In 1987, Brand co-founded the Global Business Network, a consultancy made up of 'remarkable people' including Doug Engelbart and Esalen co-founder Michael Murphy. The GBN spread its vision of the free and wired future deep into the offices and brains of the powerful, advising clients such as IBM, AT&T and the Pentagon. That vision would do much to sculpt both today's age of perfectionism and the selves living within it.

Theirs was a bubbling and urgent mix of ideas from every era we've visited. What's most clearly in evidence is the humanistic conception of people as gods, Esalen's obsession with authenticity and 'openness', the self-esteem movement's roiling narcissism and the neoliberalist's will to see competitive free markets turning the human world into a game. They began pushing the view that a new age of wired technology was going to knock down the old systems and unleash an era of unprecedented freedom and individuality. The hierarchies of government and the corporation – so collective, so coercive – would be levelled and a new civilization would emerge in which every human was their own independent hub of ingenuity and profit-making.

In 1997, GBN member Peter Schwartz co-authored an influential essay called 'The Long Boom: A History of the Future, 1980–2020' that sketched out its path. Calling his vision 'not a prediction, but a scenario, one that's both positive and plausible', he observed that the US was in the middle of a period of economic prosperity and that 'barring some bizarre catastrophe, a large portion of the world will continue to boom . . . on a scale never experienced before'. The neoliberal reforms introduced by Thatcher and Reagan which at the time looked 'brutal' had paid off. These changes, coupled with new advances in computer technology, would lead to the US being awash with riches by 2000. 'With the coming of a wired, global society, the concept of openness has never been more important. It's the linchpin that will make the new world work. In a nutshell, the key formula for the coming age is this: Open, good.

Closed, bad. Tattoo it on your forehead. Apply it to technology standards, to business strategies, to philosophies of life. It's the winning concept for individuals, for nations, for the global community in the years ahead.'

For 'open' read 'free': markets free of regulation; business free to do whatever it pleased; individual women and men 'free' to play the neoliberal game of life. On paper, many members of GBN were libertarians, which is the political ideology most often associated with Ayn Rand and her followers. In practice, libertarianism and neoliberalism have a huge amount in common, not least in their core belief in the benevolence of free markets, and their hatred of central planning and the state. Alan Greenspan called himself a libertarian, as did many of Stewart Brand's associates. *Wired Magazine*, which would do perhaps more than any other title to promote his worldview to the public, was partly funded by the GBN and co-founded by libertarian Louis Rosetto. (A 1994 edition covered the GBN itself, depicting the people in its network visually as a series of interconnected hubs – the member with the most connections of all was Esalen's Michael Murphy.)

Between them, the GBN, the Whole Earth Network and *Wired Magazine* were fantastically successful in popularizing this neoliberal vision for the digital future during the 1990s. 'By the end of the decade,' writes Turner, 'the libertarian, utopian, populist depiction of the internet could be heard echoing in the halls of Congress, the board rooms of Fortune 500 corporations, the chatrooms of cyberspace and the kitchens and living rooms of individual American investors.' *Wired* and the network of people from the Whole Earth scene 'legitimated calls for corporate deregulation, government downsizing and a turn away from government and toward the flexible factory and the global marketplace as the principal sites of social change.' A new order was coming: closed hierarchies of business and corporations would fall, government would shrivel, and the power of the open network, connected by computers, would free individuals to become wealthier than ever before. The gamification of human life would be dramatically intensified thanks to silicon technology. But, as with all great cultural movements, none of this would've taken hold in the first place if it hadn't already resonated with something deep in the selves of the people. Theirs was a conception of tomorrow that was perfectly designed for the neoliberal self that had been cooking since the 1980s and was now, in the 1990s, greasy with self-regard.

Two years, eight months and ten days following the publication of 'The Long Boom' came the crash. On 11 March 2000, investors began reacting to their realization that the tech companies they'd been raining money on had been drastically overvalued. In less than a month, nearly a trillion dollars vanished from the stock market. But once the embers of those early dot-coms had cooled, a new phase of the wired, internet age began warming beneath its dust. This re-boot, 'Web 2.0', emerged as a digital grandson of Esalen. It came enhanced by the I-focused technology of Engelbart, the neoliberal individualism of Brand's Global Business Network and not a little added narcissism. It would be a kind of global online encounter group in which the currency was the self and its gold standard was personal openness and authenticity. The future of the internet would be 'social'. It would reassert the power of the 'I'. Its platforms would flatten hierarchies further, giving every 'I' a voice, a character, a presence, a brand. It would ride on top of our increasing sense of individualism, taking the neoliberal game of life to previously unimagined places, pitching self against self in a ceaseless competition for followers, feedback and likes. But it would also tap into darker wells – ones that connected with powerful, primal instincts for status, reputation, moral outrage and tribal punishment.

In December 2006, in honour of this development, *Time Magazine* awarded their famous annual Person of the Year cover to 'You'. Two weeks later, Steve Jobs introduced the world to the iPhone. Between 2006 and 2008, Facebook's user base grew from 12 million to 150 million. Between 2007 and 2008, Twitter went from hosting 400,000 tweets per quarter to 100 million. In 2010 the first iPhone with a front-facing camera arrived. Although it was envisioned primarily as a tool for video chats, we the people surprised the technologists by mostly doing something else with it. By 2014, 93 billion selfies were being taken every day on Android phones alone. Every third photograph taken by an eighteen- to twenty-four-year-old was of themselves. 'You' had arrived. And to get along and get ahead in this new you-saturated social media arena, you had to be a better you than all the other yous that were suddenly surrounding you. You had to be more entertaining, more original, more beautiful, with more friends, have wittier lines and more righteous opinions, and you'd best be doing it looking stylish in interesting places with your breakfast healthy, delicious and beautifully lit.

In 2008, this already intensely competitive, status-obsessed neoliberal realm collided with a global economic catastrophe. In the fallout of the financial crisis, even greater pressure was placed on the ordinary Western individual. The age of perfectionism was here.

*

My train passed through Menlo Park, where John Vasconcellos had recuperated following his heart bypass operation, where Doug Engelbart had developed his astonishing vision of personal computing and where Stewart Brand's Whole Earth organization had been headquartered. It rolled through Palo Alto, home of Stanford University and Xerox PARC and then Mountain View, where Engelbart got his first job in the Valley, and which is now Googleland. The neighbourhoods were disappointingly ordinary to look at: blocks of bungalows – terracotta, dun, dusty white – with outsized SUVs in their drives. In the yards, great palm trees that had overgrown their plots gave the landscape a strangely hairy appearance. There was a Sunday evening atmosphere; recycling bins, concrete Buddhas and men in shorts with prominent veins riding expensive bicycles. Finally, the train stopped at Sunnyvale, where you alight for Cupertino, home of Apple. I took a taxi up a winding hill to the Rainbow Mansion, a large cream house with a terracotta roof that was now an 'intentional community' made up of bright young tech workers.

My search for the self had taken me, once again, to the Western world's edge. Now that I'd landed in the present day, I had the opportunity to meet selves that should be, if the ideas I'd been pursuing were correct, made up of so many of the lives of their ancestors. Being born and raised in the 1980s and 1990s, these were thoroughbred neoliberals, the kinds of people that many of the cultural leaders of our past, from Aristotle to Stewart Brand, via Ayn Rand and her Collective, should in theory recognize and think of as good. I wanted to see if I could detect their ghosts haunting the men and women who were making our future, and to discover the extent to which they'd internalized the economy of their time, fashioning it into a sense of who they were, who they wanted to be, and how the world ought to work.

Since the tech boom had drawn so many of the world's most advanced minds to this area, shared living had become extremely popular. San Francisco now had the highest rents in America – the average cost of a one-bedroom apartment was 23 per cent higher than in New York. These teaming accommodations were often known as 'Hacker Hostels', and the local newspapers had become greedy for the tales of scandal and eccentricity that seemed to beam from them like Wi-Fi. One, the Negev, had been reported in the *San Francisco Examiner* for charging $1,250 a month for a place on a bunk bed, for crowding sixty people in a space meant for twenty-two, for cockroaches, for mice and for a 'consistent odor of gas from a broken water heater'. Meanwhile, Chez JJ, in the Castro, had been closed after repeated complaints from a neighbour – a forty-six-year-old government worker – who allegedly felt sufficiently tormented that he'd taken revenge by playing a children's song on repeat and at full volume then leaving the house for hours at a time. The Startup Castle, a Tudor-style mansion in the Valley, became the subject of mockery when its entry requirements were revealed. Applicants were required to exercise fifteen hours a week, have a 'top class' degree and enjoy petting dogs. 'Dealbreakers' that would keep you out included getting regular gifts from your parents and having been 'prescribed anything by a psychiatrist more than once'.

My contact at the Rainbow Mansion was a Spanish NASA employee called Vanesa. 'You're in the Mystery Room,' she told me, leading me through the tall, shiny lobby, with its tattered chandelier and its wall calendar on which someone had scheduled 'World Domination' for next month. Vanesa was the person who dealt with the applications that were sent in, from all over the world, by workers trying to get a bed here. 'We attract high quality people,' she told me, plainly. 'The first filter is if I think they're interesting enough. We want people who have lived internationally, people who are doing amazing things in their careers and people who are easy to get along with. If they have networks, that can be really useful too, because then we have access to more people like them.'

She showed me a shared room, upstairs, where temporary guests auditioning for a place in the house would sleep. Their beds were neatly made and cluttered with belongings: the books *The Psychology of Influence and Persuasion* by Dr Robert Cialdini and *Zero to One* by Peter Thiel, the influential libertarian investor, Trump supporter and co-founder of PayPal, lay near a comedy Viking helmet and a bright red squirt gun. During their trial, the Mansion's permanent residents would have meetings in which the auditionees' potential career trajectory and personalities would be discussed, judged and voted upon. It seemed to me that in order to win a place here, the main thing you had to be was beneficial to everyone else. They were all nodes in the network, these people, individual islands of knowledge and power, only as good as the value they offered. It seemed harsh, but also rational and pragmatic . . . and it seemed neoliberal and GBN and libertarian and Greek.

That evening, as I was kicking back on my mattress in the Mystery Room, a mechanical engineer named Jeremy popped his head in. He was thirty-five, as keen as a puppy and dressed in a plain grey T-shirt and cargo pants that had various pockets and pouches hanging off them. A veteran of three start-ups, Jeremy was now working on a secret project for an extremely famous tech brand that he asked me not to name, but granted me permission to describe as 'a company that's normally known for internet search'.

We went out for burgers. As we drove into Cupertino, passing mysterious low buildings marked only with numbers and glowing Apple logos, I asked Jeremy how he'd come to be recruited by the extremely famous tech brand that's normally known for internet search. He explained that they'd approached him after hearing about a project he'd been working on.

'It was a wrist-worn wearable that had the form-factor of a watch, then unfolded to a quad-copter, deployed thrusters and propellers, flew away from you, stopped, took your picture, flew back, and then folded back into your wrist as a watch.'

'I don't believe you,' I said.

'I'll show you a video.'

'And you're telling me it knows how to stop in midair and find your face?'

'It has a whole suite of sensors to figure out where it is and how to come back.'

'Oh fuck off,' I said.

We drove for a while in silence. 'So it's like a selfie drone?' I said.

'And it's a *true* drone,' he said. 'A drone is an autonomous device that makes its own decisions. This is actually a drone.'

'And how long did it take you to build this thing?'

'Thirty days,' he said.

'Oh fuck off,' I said.

'Thirty days!'

'And what kind of hours were you working?'

'All of them,' he said. 'It was very intense. There's a reason so many of these entrepreneurs are young. Your mind and body really take a huge beating. Most people burn out. Some jobs I was sleeping in warehouses on cardboard so I could stay up just enough.'

'You were sleeping under your desk?'

'I didn't even have a desk. I was working twenty hours, some days. The timescales are short, time is money, the VC [venture capital] money is going to run out, a lot of times they pay you in equity, so if the company goes bankrupt you don't get your money. It can be a real meat-grinder.'

As we sat down with our burgers and fries at Five Guys, Jeremy showed me a YouTube video of the selfie drone in action. I quietly marvelled at how impressive all this was, but also how brutal. Had Ayn Rand and Aristotle been at Five Guys for an All-The-Way with Cajun Fries that night, they'd surely have embraced Jeremy like a son. I'd come in search of the neoliberal self and it seemed I'd found their teaming capital. But what goes in Silicon Valley, of course, is increasingly going everywhere else. This vision, of individuals 'free' to get along and get ahead by zooming unfettered from job to job, is what's often known as the 'gig economy'. It appears, too, in the guise of the 'zero hours contract' worker. They're arrangements in which the responsibility of the employer is minimized, and that of the individual maximized. It made me think of the man at the counter at Esalen who'd refused to take my bag; *we don't take that responsibility*. It was turning out to be the mantra of our times.

Back in 1981 Margaret Thatcher had said 'economics are the method but the object is to change the soul'. The truly extraordinary thing, for me, was that the soul of the people really *had* changed – and by exactly the method Thatcher proposed. Today's young people have grown up in an era in which the markets have been freed to rule, and free markets favour the self-starting business-person who relies on their own hard work and guile for survival. This generation have internalized the neoliberal economy. It's become a chief characteristic of who they are and who they want to be. 'We live in a time where market rhetoric, for better or for worse, and probably more for worse, has become almost pop cultural,' the anthropologist Professor Alice Marwick told me. 'People are interested in entrepreneurship as a general social value. They're admiring celebrities for their hustle, avidly following things like which album or movie makes the most money or who's the most highly paid. There's a sense you go after corporate money, and that way you're independent and no one can tell you anything. But what that really means is we're allowing the norms and values of corporations to dictate how we behave in our daily lives.'

In the age of perfectionism we're more likely to simply accept we shouldn't rely on the collective organism, whether it be government or corporation, to take responsibility for us. 'We've really moved away from companies taking responsibility for training employees and keeping them for years and years,' she said. 'We reward this almost constant stream of freelancers, where you're only at a company for a year or two and you're responsible for keeping your skill set up to date and if you don't do that, and you can't get a job, that's your fault. It's not the state's fault for not taking care of its citizens, it's your fault for not keeping up with what's expected of you. I think people have really bought into this idea.'

But when an economy like this becomes a part of who we are, it can be dangerous. It means we tend to judge people who actually need help harshly – and judge ourselves harshly when we need it. We decide we're failures. We're losers in the game. And if we tend towards perfectionism, and are more sensitive to signals of failure in our environment, we risk the self becoming unstable, bringing thoughts of suicide and self-harm.

Part of neoliberalism's genius is that it has, as its electricity, our natural desire for status – it rewards the impulse for getting ahead of the rest of the tribe that's inherent in the human animal. Just as its founding father, Freidrich Hayek, had planned, this means it doesn't have to be imposed by force, as failed ideologies such as communism and fascism did. 'The market manages to regulate the people by having them regulate themselves,' Marwick told me, 'so the state can just sit back and watch.'

Some of the effects that this post-self-esteem, neoliberal economy has had on our values have been charted by the narcissism experts, Professors Twenge and Campbell. They cite surveys that suggest marked changes in values, among the young, especially in areas such as individualism and materialism. In 1965, for example, 45 per cent of freshmen said they believed it was important to be financially well off, a figure that had risen as high as 74 per cent by 2004. High-school children in the 1970s were half as likely as those in the 1990s to believe 'having lots of money' was very important. 'The biggest change in culture, from the baby boom to today, has been this thing about finding a meaningful philosophy of life,' Campbell told me. 'And the biggest increase is wanting to make more money. There's just a big change from internal to more external values.'

All of this finds expression too in the nature of our ambitions. 'I mean, you guys in the UK have this fame culture. We got a lot of it from you.' One 2006 poll of British children placed 'being a celebrity' at the top of their list of the 'very best things in the world'. Over in the US, that same year, another survey found just over half of eighteen- to twenty-five-year-olds naming 'becoming famous' as a generational priority. Twenge points to further data that suggest individualism is rising, including greater use of self-focused language in books and song lyrics and the increasing popularity of unusual baby names. 'But one of my favourite pieces of data is the amount of people in our country having the mystical experience of a direct connection with God,' said Campbell. 'It's doubled since the 1960s.'

In their book, *The Narcissism Epidemic*, Twenge and Campbell describe a further cultural change that I thought especially remarkable, given my experiences at Pluscarden Abbey. They write of the largest place of worship in America, Lakewood Church in Houston, which apparently has 'perfectly airbrushed' portraits of its pastor, Joel Osteen, lining its walls. 'God would not have put the dream in your heart if He had not already given you everything you need to fulfil it,' Osteen has written, going on to explain that 'God didn't create you to be average. You were made to excel.' It was hard to believe this was the same religion as that which I'd immersed myself in, back in Scotland. We'd come a long way from 'I am truly a worm.' Neoliberalism and its self-esteem revolution had even, it seemed, helped redesign God.

When we're bathed, from birth, in our particular economy, its values start to leak out in the things we say and do. Neoliberalism beams at us from many corners of our culture and we absorb it back into ourselves like radiation. It's in our television programmes and on our supermarket shelves; it's even in our prescriptions. 'If you look at things like makeover shows, or weight-loss products, or self-help books about becoming more productive, or ADHD drugs that help you work longer hours, it's all the sense that you're taking on personal responsibility for becoming a better employee or becoming a better person,' said Professor Marwick. 'And if you're becoming a better person, that often requires buying many things or regulating your body according to what a proper neoliberal citizen should be – someone who's healthy and active and participating and a self-starter.' On the upside Marwick believes it's been a factor in our increasingly progressive stance on minority issues. 'The big push behind gay rights and gay marriage is a neoliberal product,' she said. 'Gays and lesbians are great consumers. So it's not all bad, is what I'm saying.'

*

As we were finishing our burgers, I asked Jeremy if he had any idea why conspicuous shows of wealth seemed to be so unfashionable in Silicon Valley. This felt like an interesting cultural quirk and I wasn't sure what to make of it. 'Well, it's because it's very easy to get a huge amount of money very, very fast and people are not impressed by it,' he said. 'It's much more important to be known for what you've made.'

'But why?'

'It's another facet of power, to have something you created affect the lives of millions of people. In Silicon Valley, we're all trying to change the world. There are people who say that and there are people who believe it. In my experience, if you're running solely on one aspect, you're going to have trouble. If you're just looking for money, or you're just looking to change the world, you're going to be all out of whack.'

'Why?'

'Because if you don't change the world and you don't make a lot of money, you're a failure.'

As we drove back to the Rainbow Mansion, I told Jeremy that there was a particular species of tech worker I was especially interested in tracking down. Just as the culture in which we're suspended forms a generalized 'perfect self' for us to compare ourselves to, so every cell of that culture also generates its own specialized iteration of it. In 1943 Jean-Paul Sartre noticed that there was a cultural concept of the perfect waiter just as there are models of the perfect teacher, the perfect pop star, the perfect politician, the perfect parent and the perfect boss. In Silicon Valley there's a model of ideal self that many younger tech workers aspire to be that's become significantly influential throughout the West. It's a concentration of all their hopes, values and ideals. It is 'The Founder', the plucky self-starting genius, who through inspiration, bravery and hard work has not only made themselves rich but who has made the world a better place.

When I asked Professor Marwick, who carried out extensive fieldwork in Silicon Valley during the Web 2.0 era, to describe this strange archetype, this is what she told me: 'The Founder is visionary. The Founder is intelligent. The Founder is independent. The Founder is iconoclastic. He is usually a white man under thirty who went to Stanford or Harvard. He wears hoodies. He wears sneakers. He has a bike. He is a very hard worker. He is uncompromising in his worldview. He has some sort of vision he's going to execute and he's going to attempt that, regardless of what gets in his way. He's nerdy, a little strange, and has some personal quirks. But he's tech-savvy, always on, always connected. And he's going to change the world.'

'Oh, you should speak to Dan,' said Jeremy, when I asked him. 'He's the one running Deep Space Industries.'

'What's Deep Space Industries?' I asked.

'They're trying to mine asteroids.'

'Trying to mine *asteroids*?' I said.

'Yup.'

'Oh fuck off.'

Back in bed, in the Mystery Room, I sat up on my pillows, and picked through my notebook. Before my arrival in Cupertino, I'd spent some time at a Hacker Hostel up in San Francisco. It was called 20Mission and was popular enough that its regular parties were reported in style magazines and typically attracted more than eight hundred fashionable guests. Back there I'd found yet more evidence of our modern absorption of the ideas I'd been tracking since 500 BC.

You entered 20Mission through an unmarked door behind a black metal grille that released its lock via a smartphone app. The walls of its four corridors, arranged in a square, were painted with 1960s-style hippy motifs and cutouts of butterflies and snowflakes. Off the corridors led forty-one heavily stickered bedroom doors ('I ♥ Robots'; 'Die Techie Scum'). In the centre was a courtyard in which they'd kept chickens until the city council found out and forced them to pay for retrospective architectural plans for the coop. The whole project ended up costing $10,000. The chickens supplied one egg. When 'House Mom' Diana gave me a brief guided tour, she told me there'd been complaints of supernatural disturbances. 'I believe there's a hooker ghost,' she said, flatly. 'It's a woman who got murdered here.'

'And what does the ghost do?' I asked.

She looked at me as if I was stupid. 'Well, she's a hooker,' she said. 'So she rapes you.'

In a different time, the building that 20Mission occupied had been the Sierra Hotel. There was currently huge demand for places like this – converted SROs, or 'single room occupancies', that used to be known as fearsome places full of itinerant workers who would drift from job to job but are now known as desirable places full of itinerant workers who drift from job to job. 'People's work lives are very fluid, especially with software engineers,' a young entrepreneur called Shannin told me, in the kitchen that morning. 'I don't know a single person who's like, "I'll just get a job at Google; I see myself there for fifteen years." It's like, "Yeah, I see myself here for a year or two and we'll see what happens after that."'

When he'd finished preparing his lunchtime broccoli for baking in the oven, I followed Shannin down the corridor to his room, which was as smart and orderly as he was: books filed neatly, drawers partitioned into little sections, a spotless $3,000 carbon-frame road-bike hanging on the wall. He was as beautiful as a choirboy and thin as a pen. He offered me a slice of the sourdough he'd baked that morning. It was wrapped in a little white sheet. It was delicious. He'd moved to San Francisco two years ago, at the age of twenty-two, and worked as a software engineer for a couple of start-ups, but had since decided to pursue the ultimate dream. He'd ventured out alone, into the neoliberal wilds, and was aiming to become a successful founder. When he was scoping for a direction in which to launch his ambitions, friends told him, 'You like to bake and you like to get baked so you might as well make edibles.' He'd decided to produce chocolate oranges soused in cannabis oil and sell them for $25. 'Last year, in California alone, there was over one billion dollars in legal cannabis sales,' he told me, with great seriousness, as if I was a VC about to reach into my wallet and purchase 3 per cent of his business.

When he told me he preferred to work for himself, I asked why. 'For a while, a good life was, you got a good education at a top-tier university, then you go on to one of the big companies, like a GM or Ford, a big institution, and you work your way up that ladder, and that was what you did. Then people started realizing that's not the only way to live. Everybody started focusing more on individualism. It's gone from working for the company, where you've got the secure job and you get the car and the health insurance, to focusing on what you're actually, truly passionate about. Something that you actually have your hands in rather than just doing it for somebody else and getting that paycheck in return.'

I asked him to describe his vision of the Founder. 'The Founder is somebody who sees their product as their child,' he said.

'So it's having a vision beyond money?' I asked.

'That's what a lot of VCs look for,' he nodded.

I left Shannin to cook his next batch of edibles and returned to the kitchen, where I found Berkeley, an earnest twenty-six-year-old in a plaid shirt with cropped hair and a neat beard. Berkeley helped run 20Mission. He'd made his money doing something with the virtual currency Bitcoin that I pretended to understand but didn't. We began talking about the role of government, and collective projects for the common good, and it quickly became clear he was deeply sceptical. 'It's kind of a weird thing, when people say we need to do things for the common good,' he said. 'Tech companies made it so that I could get a ride anywhere in the city for five dollars, door to door, and split it with two people. San Francisco would be impossible without Uber and Lyft. And then the city come here and fine me because people put graffiti on my windows.'

'They fine *you*?' I said. I could see why that would be annoying.

'There are times where some of us have said, "Can you imagine where we did an experiment where Google took over what City Hall does now?"' he said. 'Just let them have at it and see what they can come up with?'

It seemed clear Berkeley thought that would be an exciting idea, that tech brains were of a different order to the ones the politicians and civil servants carried around in them. I asked about major Great Compression projects like the New Deal. Weren't those noble undertakings for the common good of the kind that could only be provided by a collective organization such as the state?

'Those ideas have been tried and some would say they didn't work out as planned,' he said, with a smile.

'In what way didn't they work?'

'Well, we had a so-called war on poverty in the US. Poverty is still there.'

'But isn't it also simply about giving people on the bottom of the socio-economic pile a basic standard of living? There doesn't have to be an end goal beyond making sure they're not completely miserable.'

'But we still have that, to some extent.'

He seemed functionally unable to accept what I was saying, as if the continuing existence of poverty and misery was an argument against trying to altruistically lessen their effects. It was the kind of mindset that might be baffled by the idea of a National Health Service, for example, or of state funding of the arts; the kind that could only understand the logic of the game and its ruthlessly competitive sorting of winners and losers.

It made me wonder how he identified politically. I'd heard rumours Ayn Rand and the libertarian ideology she's associated with were popular around here. Steve Jobs, for example, is said to have treated *Atlas Shrugged* as his 'guide in life', whilst former Uber CEO Travis Kalanick used the cover of *The Fountainhead* as his Twitter avatar. But, for some reason, it's rare for tech workers to publicly declare love for the author of such works as *The Virtue of Selfishness.* Take one widely publicized 2017 survey of tech leaders' political beliefs by Stanford University researchers. It found strong disapproval of labour unions and state regulation of business, as well as a widespread belief that protections for workers are too robust and bosses should be able to fire them more easily – so far, so Ayn. But when asked outright, only 24 per cent were prepared to admit endorsing the libertarian philosophy in full. Oh no, they insisted, they were actually in favour of increased taxation, redistribution of wealth and state care of the needy.

But the tech leaders' gentle claims stand starkly at odds with their real-world behaviour. They use their considerable lobbying might, not to press for their fellow rich to be forced to pay up for a more benevolent state, but to fight regulation and erode workers' rights, thereby increasing the profitability of their companies and, by extension, their personal fortunes. Plus, you might think, if they really *did* believe in redistribution of wealth, state care of the poor and strong taxation, there's one simple thing they could do immediately that would fulfil all their most altruistic, anti-Randian, pro-tax fantasies – they could actually pay their fair share of taxes.

I wondered where Berkeley stood on all this. 'I'm libertarian-ish,' he said carefully. 'I don't really subscribe to one ideology, but it does seem . . .' He stopped himself. Soon, though, he began talking about start-ups as 'creative' acts. The logic seemed to go something like this: just as in the 1960s and 1970s people were encouraged to actualize through self-expression – singing, painting, poetry, fashion and so on – their modern descendants are actualizing through their businesses. 'Starting your own company, versus art. Where do you draw the line?' he said. 'We express ourselves with what we do with our lives. For me, I'm not a good painter, but there are things I do that are artistic.' It was a fascinating idea. For one of us our core personal project could be writing an opera but for another it could be founding a company. I could think of someone else who might have appreciated Berkeley's logic.

'Have you ever read any Ayn Rand?' I asked.

'Only *Atlas Shrugged*,' he said, shrugging himself.

'It's just that, what you were saying about your business being an act of creation, that sounds like something she'd have appreciated.'

'That's interesting,' he said. 'I hadn't thought of that. I did enjoy a lot of things about *Atlas Shrugged*.'

'I heard she's very influential in tech circles,' I said. There was a silence. 'I don't know.' He turned to a man with long hair who'd walked in and started baking broccoli. 'What do you think, Chris?'

'There's sort of this idea that her books are into a meritocracy and that's the one thing the tech community really likes,' he said, pushing his baking tray into the oven. 'I did enjoy *Atlas Shrugged*, but it's weird, a lot of people will think very bad things about you immediately if you say that.'

Berkeley explained that he used to be more liberal but found his mind inching rightwards when he spent a period working overseas. 'When I grew up, I was told there was this evil dictator in Latin America that the US put there and everyone hates us for it,' he said. 'But it turns out that I go there and people don't really hate the US and a lot of them are, like, "Yeah, that dictator, he wasn't the greatest guy, but now we actually *have* things."'

'And who was that dictator?' I asked.

'Pinochet.'

I took myself off to the communal lounge where I sat in a deep, low armchair. There were men with shiny skin smoking pot and playing *Fallout* on a sixty-inch television, an electric guitar propped up in the corner, empty wine bottles, dropped playing cards and, on the wall, a large purple throw depicting a meditating Buddha. The shades were drawn, the air was smoky, time was slowed. Despite never having been in the room before, I recognized it instantly from my own twenties, when I'd smoked skunk and watched people play PlayStation for hours. I felt awkward. I felt old.

'So where do you come from?' one of the *Fallout* boys asked me.

'UK,' I said. 'About an hour out of London.'

'Whereabouts out of London?'

'It's a place called Kent.'

He took a deep draw of his marijuana pipe and studied me with heavy lids, his bare feet stretched under the coffee table.

'What do you, like, do in Kent?'

'Um, I like going on long walks, really,' I said. 'And that sort of thing.'

He began nodding very slowly, his eyes fixed on my face.

'Sweet. Cool. That's awesome.'

On the computer, gunfire blasted.

I sat for a while, faking a long, fascinated squint at the bookshelf, before trying to restart the conversation. No words arrived in my mouth. There was a horrible possibility, I realized, that I was getting stoned. I smiled, palely, at a woman with red hair and a tiny fringe who was concentrating heavily on her colouring book, blue pencil in hand. She turned out to be Stephanie Pakrul, aka StephTheGeek, who was described on her own Wikipedia page as an 'internet personality'. Today, it turned out, she'd eaten nothing but the synthetic food-replacement drink Soylent. Later, she'd put some medical marijuana in a batch and tweet about it being 'the most San Francisco thing ever'. Her tweet would be retweeted by the Deputy Technology Editor of the *New York Times*.

Steph blogged on Medium, posted on Tumblr and made money as a web developer and as a cam girl. She was, according to her profile, turned on by flattery and getting tied up and was an expert in debugging code and 'squirming'. I'd heard rumours that polyamory was fashionable in the tech scene, monogamy perhaps being seen as yet another antiquated technology ripe for disruption. 'Half the people here are poly,' she told me. 'And there's an orgy culture here, definitely. There are people here who have very active sexuality and expressiveness, have multiple partners, have a lot of kink going on, go to sex parties, go to BDSM parties and all that kind of stuff. Like, major, major, like, open-sex people. It just permeates.'

What seemed to connect the sex and the start-ups was a commitment to openness and authenticity as well as an implicit faith in libertarian-neoliberal individualism: that putting the self's desires above all will always lead to progress. 'It's a great anti-Luddite kind of thing,' she nodded. 'It's about "let's drive the world forward, let's use all this technology but let's also stay connected to each other and, like, all the hippy-dippy wonderfulness". And then there's the whole, like, Singularity.'

'I've never really got my head around exactly what the Singularity is supposed to be,' I said.

'It's basically that we're going to keep making more intelligent, faster and faster machines and all of a sudden, there's going to be this exponential, just like, brrccccchhhhhxxxsss-zzzzz, and we don't know what happens. Some people say the Singularity is, like, the world's end, some people say we're going to evolve into another species. It's meant to represent acceleration, change, and then a moment of complete transformation of the universe.'

I had no idea what she was talking about.

I left, that night, in a daze, my head heavy with visions of the brrccccchhhhhxxxssszzzzz future and with medical marijuana. The singularity, polyamory, roasted broccoli, Pinochet . . . it was enough. I thought I'd clear my head by walking back from the Mission district, where 20Mission is based, to my downtown hotel. About forty minutes into my journey, I entered an area of Market Street that looked as if it had been set-designed by the Devil. Hanging in the glow of a twenty-four-hour liquor store was a scattered wolf-pack of ghouls, hustlers and the deranged. A middle-aged man staggered into the road with his trousers and underpants around his knees, his testicles low, his thighs soiled with faeces. I crossed the street, worried about offending them and worried about my wallet. I quickened my pace. To my right I saw the smartly glowing headquarters of Twitter, to my left, the proud dome of City Hall.

*

The next morning, just before breakfast, I met him. He was cooking pancakes in knee-length blue running shorts with his top off, the thirty-eight-year-old CEO of Space Industries, Daniel Faber, a tall, athletic, charismatic genius from Tasmania. Sat on a stool at the breakfast counter, I enquired after his path into asteroid mining. 'I went to the University of New South Wales and eventually got bored of hang-gliding, windsurfing and building solar cars,' he told me. 'I decided that to benefit the most people in the world I'd have to address existential risk. I thought about that for a bit and realized getting people off this rock was the first thing we should do. So I made a list of all the things that could pay for the first permanent job in orbit.' There were three items on his list: space-based solar power, space tourism and asteroid mining. 'Solar power needs resources in space otherwise it's not economic and I couldn't see myself as a tour operator. So here we are.'

'So you're actually . . .' I was trying to find a way of expressing my incredulity without appearing rude. 'You *actually* think this is going to happen?' I'm not sure I was quite managing it. 'You're *actually* going to mine an asteroid?'

'Nobody doubts this will happen,' he said. 'The only question is when.'

He was cooking two pancakes in two pans simultaneously. Sweat glistened on his shifting biceps. He was oozing *kalokagathia* and looked as Greekly perfect as that sexy Jesus I'd seen, all those months ago, hanging above my bed at Pluscarden.

'And what are you going to mine?'

'We're not trying to bring material back to Earth. We're after hydrocarbons, water, nickel, iron. All the materials you'd need to build cities in space.'

'And you *actually* think we'll live to see people living in space?'

'I think I'll live another thirty years, yeah,' he said. 'Elon Musk wants to put people on Mars by 2026. Anyone else at any other time in history would've been mad to say that. But this is Elon Musk.'

I wondered about the influence of Ayn Rand among his fellow founders. 'Engineers and richer folk are often libertarian,' he said. 'It's never been tried, this pure libertarianism that Ayn Rand was promoting. What we need is a chance to give it a go. If we had a whole bunch of habitats in space that were somewhat politically isolated, you could run these experiments. That's one of the things that will inevitably happen. Someone rich enough will build one of these places and go and do whatever they like.' I asked him why libertarianism seemed to appeal to engineers. 'I think they believe they can design something better,' he said. 'Given all the flaws that we see in the current system, I can see, absolutely, that that's attractive.'

'And you?'

'I'd lean towards libertarianism,' he said. 'I'll wear my colours. I also believe there'll be unintended consequences to anything that gets set up. But without trying it, we won't be able to figure out the solutions.'

We sat down and ate Dan's pancakes.

They were perfect.

*

2016 saw the first rumblings of mass rebellion against some of the effects of neoliberalism. In classic story-style, low-status forces 'below the line' came together in an attempt to topple those 'above' in campaigns of such power they surprised even those who led them. Tracing what happened in that tumultuous year means stepping back into the collective 1960s – the civil rights era during which the left began turning its focus towards equality and minority issues. It was then that, feeling ignored and resentful (and, in many cases, racist), a number of working-class Democrats deserted the party for the Republicans. In 1964, 55 per cent of all working-class voters were Democrats. By 1980, that number had fallen to 35 per cent.

Under the inequalities of neoliberalism, the white working class suffered. The new era of globalization it brought about saw some of the manufacturing and service industries they relied upon moving overseas. Many others lost their jobs because of automation, the effects of which a more collectively minded state might have sought to mitigate. Whilst plenty of people have become better off, since the 1970s, a good deal of others have seen the worth of their paychecks stall or fall. The average real income for the bottom 90 per cent of earners in the US, for example, has pretty much stagnated. It was $35,411 in 1972, it peaked at $37,053 in 2000 and, by 2013, had fallen to $31,652. Whilst those numbers hide rises for the well educated, they also obscure falls for the less. During the Great Compression, those without a college education were able to work in a unionized job, own a car or two, and live in a suburban house with a backyard. But, for many, those days became a memory. Between 1979 and 2005, the average real hourly wage for those without a high-school diploma, a group that shifted significantly towards the Republicans in the 2016 election, declined by 18 per cent. One analysis found that the slower the job growth, in any given county, the more it swung towards Trump.

Political scientist Professor Katherine Cramer spent several years amongst the rural working class of Wisconsin, a state that turned Republican, in 2016, for the first time since 1984. She describes a powerful sense amongst the population that over the last four decades something had been going badly wrong; that they were working as hard as their parents had and paying all their taxes and yet being rewarded with a much poorer quality of life. 'They feel like they're doing what they were told they need to do to get ahead. And somehow it's not enough.' Despite their toil, they weren't receiving any payback in the form of power, money or respect. Their sense of efficacy over their lives and the wider world had been halted, their heroic plots stopped. This created a deep grievance that curdled into a generalized disdain for the people they held responsible: the elites, the city people, the establishment, the government. In their minds, they were all tied together in one hopeless, heartless, arrogant, corrupt tangle.

Encouraged by messaging from the Republicans and right-wing media, many decided the answer lay in getting rid of as much of that hated government as possible. The year of the financial crisis saw the beginning of a sharp and sustained upturn in support for smaller government. And it wasn't just the poorest that were behind it. More and more Americans began agreeing with statements such as 'government has gotten involved in things that people should do for themselves' and 'the less government the better'.

The 2000s also saw a rise in popularity of the idea that governments should be run more like businesses, a trend that starkly exposes the mass internalization of the neoliberal self. The arrival of this notion has not only been tracked by political scientists, it was found by Cramer in her Wisconsin fieldwork. 'This came up often,' she writes, 'and not just in predominantly Republican groups.' In Donald Trump, of course, millions were to find their businessman. Here was a man who viewed the world as a system of transactions, of profit and loss, of deals won or squandered. As well as this, he gave voice to their feelings of betrayal and promised to 'drain the swamp' in Washington. He would make them heroes once more.

They might not have known exactly what to call it, but many of those voters were right in their hunch that *something* had changed a few decades ago, and that they were suffering because of it. They were also justified in feeling ignored. Since the end of the Great Compression, voters at the ballot box had really been given a choice between two different forms of neoliberalism. Is it any wonder that so many concluded, even if misguidedly, that all politicians were the same? That no matter who you voted for, nothing really changed? When political scientists compare the electorate's preferences with actual policy outcomes, they find both Democrats and Republicans tend to act mostly upon the wishes of the wealthiest, with the Democrats only marginally better at serving the middle class. The views of the poorest third of society are barely reflected at all.

It was in this context, and in the smoking aftermath of the financial crisis, that angry and frustrated people turned to anti-establishment rebels on the left (Sanders and Corbyn) and the right (Trump and Farage). True to our individualist core, the socialists faltered. And so came the political shocks of 2016. In the UK, the vote to leave the European Union was driven by a campaign that played especially on the fears of those discomforted by the free movement of labour, with alarmist posters warning, 'TURKEY (population 76 million) IS JOINING THE EU' and promising 'Brexit' would allow the government to 'take back control' of the borders. Meanwhile, in the US, Donald Trump said he'd 'put America first' by building a wall at the Mexican border, spending billions on infrastructure and forcing Apple to relocate their manufacturing to the US, while one of his closest advisers complained to reporters, 'The globalists gutted the American working class and created a middle class in Asia.'

To many on the left, immersed in the shibboleths of identity politics, these were outrageous appeals to old-fashioned racism, so obviously abhorrent that surely no sensible person would be able to look past them. And yet different ears heard a different story. They heard change-making outsiders scorning the establishment. They heard brave rebels disparaging the smug and 'politically correct' educated class who routinely ignored them in favour of minorities, and then patronized and insulted them when they complained. (One Trump voter's sign: 'I'M NOT "DEPLORABLE" I'M JUST A HARD WORKING, TAX PAYING AMERICAN!') As well as this, they heard arguments against globalization. They heard arguments against neoliberalism.

Which is not to say that great swathes of voters for Brexit or Trump were necessarily chatting to each other about 'neoliberalism'. Neither is it to make any comment about the truth of these politicians' beliefs and policies, and the actual effects, on the voters, of leaving the EU or shrinking the state and cutting regulations yet further. This wasn't about being up on the economic theory or parsing the detail of campaign speeches or pondering the puzzle of the poor voting for a man who'd likely remove even more of their protections. The brain, after all, isn't especially interested in facts and data. We live our lives in story mode. Our minds make sense of the world using simplistic observations of cause and effect. They confabulate, weaving a useful, makes-sense narrative out of what they're feeling and seeing that casts their owners in a heroic light. As well as this, they're often tribal, making automatic allies of those who look and think like them, and enemies of those who don't.

So what, roughly, was the story of the world that many of those 2016 voters told themselves? They had a powerful sense that things weren't as good as they had been in their parents' day, despite the fact they worked hard and paid their taxes; they knew it was the fault of the politicians, who were corrupt, with their expense accounts and incestuous connections to lobbyists and bankers; they knew that no matter who they voted for, it never seemed to make a difference; they knew that if Trump got in, there might be an Apple factory opening up some time; they knew that if we stayed in the EU a million Turks would come a-swarming; they knew that last month, when their little girl was sick, they'd waited six hours at A&E and there was a large family in front of them who didn't even speak English and how was *that* fair?; they knew the opioid addict two doors down lived on food stamps and shopped in Macy's, and that those food stamps had been paid for by their own hard toil and time; they knew the left didn't care for them any more, that they were over-educated elites who hated working folk, that the only people they gave a crap about were the minorities who were given all the handouts and were able to cut in line in front of them . . . and they knew that here, on the TV and on the internet, was a straightforward, no-nonsense businessman, who kind of looked like them, and kind of sounded like them, and seemed to understand their problems. He was promising to restart their stories, to restore the forward motion they'd lost so many years ago.

Of course, not every Trump and Brexit voter thought this way. Not all of them were intolerant of minorities, or saw the world this simplistically. No doubt, many cast their ballots holding their nose. I should also acknowledge that as a typically faulty human I can't help but filter these events through my own biased perspective. What's important, though, is the understanding that voting decisions are rarely about coldly rational economic calculations, or simple racism or misogyny. They're much more about self and story; about the tale we tell of the world, and our own heroic place within it.

Silicon Valley played its part in all this. The GBN were right in their predictions that digital technology would transform the economy. It did, facilitating and accelerating the neoliberal project of globalization immensely from the 1990s onwards. It has plenty in store for the future, too, with automation and artificial intelligence predicted to further decimate middle- and working-class jobs. There are 1.7 million truck drivers in the US alone whose livelihoods are at risk from the introduction of autonomous vehicles. Researchers at the University of Oxford have predicted that, by 2033, nearly half of all US jobs could be automated. The technologists promised us a 'Long Boom'. They didn't tell us that boom would be directed mostly at the top.

While we don't fully understand what the consequences of all this inequality will be, we appear to have entered a dramatic and bellicose period of change. On the extremes, fascism is remerging in remixed form as the racist Alt Right, whilst Marxist ideas are reawakening in the guise of the tribal, intolerant and increasingly violent Alt Left. Meanwhile, 2017 saw yet another election shock, led by UK's leftist Labour leader Jeremy Corbyn. Labour (whose shadow chancellor, John McDonnell, lists as a hobby, 'generally fermenting the overthrow of capitalism') was expected to be pummelled on polling day. But they humiliated the Conservatives, forcing them into a hung parliament and leaving Prime Minister Theresa May in shock and tears.

It might seem perverse to link the rise of Corbyn to that of Trump, but what connects them is a shared understanding that the neoliberal era is coming to an end. A voter with a right-wing brain looking at our economic travails is likely to instinctively blame immigration and welfare blaggers. A voter with a left-wing brain will look at the same mess and see the consequences of shrinking government and an under-regulated corporate-banking complex. They may have wholly opposite perspectives, but both left and right have identified the same problem – neoliberalism and its project of globalization.

During that 2017 election, even the Conservatives smelled the change in the wind. The party of Thatcher shocked many by announcing a rejection of 'the cult of selfish individualism' and a belief in 'untrammelled free markets'. In October 2017 they even promised to start building social housing. A respected survey the same year confirmed their instincts. It found 'a left-ward tilt' toward taxing and spending, support for which had risen from 32 per cent in 2010 to 48 per cent.

Silicon Valley technology played a critical role in the rise of both Corbyn and Trump. In the 2016 election, social media enabled Trump to connect directly with his supporters, bypassing traditional journalists and undermining their reporting by calling them liars. It's true that some of his strongest support came from white males without a college degree, who are among those least likely to actually use the internet. But it's also true that Trump's tweets rebounded straight back into the mainstream newspapers and television channels. In this way, he could use the internet to get his message out even to people who didn't have the internet. He could lead the news agenda, thereby continually re-affirming his status as the punk outsider, happy to fire phlegm at the establishment's chin.

In the US, polarization between left and right has been increasing inexorably since the 1970s. But tech platforms such as Twitter and Facebook appear to be making what was already a serious problem worse. Studies suggest that experiences of moral outrage in normal 'offline' life are relatively rare, with less than 5 per cent of us experiencing it on a daily basis. In the 2010s social media began drenching the general population in levels of outrage the human animal has never before experienced and is not adapted to. If that wasn't worrying enough, psychologists also believe that hearing about immoral acts online actually evokes more outrage than when encountered in person. It's too early to speculate on the full consequences of these radical changes in the public discourse. But consequences there will surely be.

Already, every day, millions of us are needled and outraged by the hysterically stated views of those with whom we don't agree. Our irritation pushes us into a place of fiercer opposition. The more emotional we become, the less rational we become, the less able to properly reason. In an attempt to quieten the stress, we begin muting, blocking, de-friending and unfollowing. And we're in an echo chamber now, shielded from diverse perspectives that might otherwise have made us wiser and more empathetic and open. Safe in the digital cocoon we've constructed, surrounded by voices who flatter us with agreement, we become yet more convinced of our essential rightness, and so pushed even further away from our opponents, who by now seem practically evil in their bloody-minded wrongness.

The effects of social media's echo chamber are magnified by invented news pieces that circulate widely on sites such as Facebook. One investigation found that in the final three months of the 2016 election, the twenty most popular false stories ('Pope Francis Shocks World, Endorses Donald Trump for President!') had more engagement with people, in the form of shares, reactions and comments, than the top twenty stories from respected sites. When it was all over, the psychologist Professor Jonathan Haidt told reporters he'd come to believe social media is 'one of our biggest problems. So long as we are all immersed in a constant stream of unbelievable outrages perpetrated by the other side, I don't see how we can ever trust each other and work together again.'

*

On my final day in Silicon Valley, I fell into conversation with Cate Levey, who worked at a start-up developing vegetarian meat. As a Christian, raised in the south, she told me she'd sometimes felt something of an outsider in Silicon Valley. It began at her first shared house in Palo Alto. From the beginning, she'd found the whole set-up of the place odd. Then she heard about a previous occupant who'd fallen pregnant with twins, in contravention of the house rule that said residents could only have one child. The group got together to have a 'rational' discussion about the problem. The prospective mother found out it was medically possible to have one of the twins aborted. This potential solution was put to a vote. Thankfully, it wasn't carried. But the fact of it happening at all was enough to send Cate looking for alternative accommodation. In its cold-blooded commitment to pure rationalism, there was something about this anecdote that reminded me of the discussion Ayn Rand's Collective had had, on whether it would be 'rational' to have Nathaniel Branden killed. What was good for the self seemed to have cruel dominion over what was good for the other.

Cate told me about a Rainbow Mansion tradition in which an interesting question is raised over Sunday-night dinner for debate. She had many friends at the Mansion and hung out there most weekends but had nevertheless felt unnerved at their answers to one particular question, 'If you could live any number of years, how long would you choose?'

'Everyone answered infinity,' she told me.

'I don't think I'd want to live for ever,' I said.

'Me neither.'

'Does it say something about their insane levels of ambition?'

She nodded. 'But also that they don't think about the rippling effects beyond themselves. They're genuinely trying to change the world but they're not thinking about systems, and how this new thing might disrupt the old world in a bad way.' Cate remembered an excited discussion at a New Year's Eve party about a cheap robot that could prepare all your food from scratch. She'd protested, offended at the potential for this creation leading to a new wave of mass unemployment. 'But all the middle-class jobs have already disappeared in this country!' she'd said. 'Not everyone can go to college and be a programmer.' Nobody agreed with her. They were making that Freudian mistake, underestimating how different they were to other people, who might be less well equipped to thrive in this hard neoliberal reality than them. And, besides, it was progress, wasn't it? And how could progress be anything but good?

'There's little compassion for people outside their immediate circles,' she said.

'It's kind of heartless.'

'Yeah.' She shook her head. 'It's making me more and more uncomfortable the longer I'm here. When you're in an echo chamber with a bunch of people who are fairly similar to you, you start saying things that would not be socially acceptable elsewhere. Like, "All religious people are idiots." I hear that a lot.'

As she was speaking, Jeremy walked past. 'They are!' he said, with a grin.

Cate waited until he'd gone. 'I just don't know what to do with myself.'

What Cate had noticed was a symptom of the hard form of individualism that characterized not only these people, but so many of us who share their culture. When we defined ourselves, all those centuries ago, as things that were separate from our environment and from each other, we turned our back on a truth that the descendants of Confucius knew well. We're connected. We're a highly social species. Almost everything we do impacts on someone else, in one way or another. Changes we make to our environment form ripples that spread out, far into the human universe. These ripples are easy to ignore. Especially for us Westerners, many are invisible. But they're there, no matter how convenient or seductive it might be to pretend otherwise and deny responsibility for anyone but our own sacred selves.

When I returned from the US, I decided I wanted to meet a person who'd been affected by some of the ripples that had come out of California. Ideally, it would be a twenty-something, a millennial, a member of the selfie generation. It would be a young man or woman who'd perhaps been altered in some way by the self-esteem movement and the ego-gilding parenting style it encouraged. And it would be someone whose narrative identity had been formed around the version of self that had come out of Silicon Valley's Web 2.0 phase. It would, in a way, be the culmination of my entire journey – the destination, in the form of a living human, to which the story of the individual, perfectible, self-regarding, godlike 'I' had been heading since we first met it on the shores of the Aegean.

*

Before raising the camera, to take one of today's many photos of herself, CJ instinctively touched her fringe, exploring it and patting it with expert fingers.

'Can you tell what your hair looks like just by feeling it?' I asked.

There was a silence.

'Nobody has ever observed me like this before,' she said. 'It's weird.'

I suddenly realized I was a middle-aged man squinting at a pretty twenty-two-year-old in a park. 'Sorry,' I mumbled.

'Don't apologize,' she smiled. 'It's good.'

I believed her. She wasn't being polite. CJ liked to be looked at. She liked it more than you can possibly imagine, at least if the selfie habit she claims to have is anything to go by. She says she's sometimes up until 4 a.m. editing, adding filters and selecting only the very finest to post on Facebook and Instagram, alongside captions such as 'Hypnotising, mesmerising me'.

'Do you keep them on loads of memory cards or delete as you go?' I asked.

'I have memory cards,' she said. 'And a one terabyte hard drive and, a couple of weeks ago, I had to buy more storage on my iCloud. I'm paying £35 a month now.'

'And that's all selfies?'

'Pretty much.'

She snapped a picture and examined it unhappily. 'I'm going to make it look a lot colder than it actually is because I look tired and I haven't got any make-up on. I have to justify why I look crap.' For the next selfie, she hugged her shoulders to her jaw and squinted her eyes in an expression of cosy dreaminess, as if smelling an imaginary hot chocolate.

'That's a good one,' I said, when I saw the result. 'Give it a mark out of ten.'

'Four,' she sniffed, getting up. 'I won't be happy until I'm at least eight or nine.'

CJ ran me through a typical day. She wakes every morning at seven thirty, thinking about what she's going to do with her hair and make-up. 'But you're thinking specifically about how it'll look in pictures?' I said.

'Yeah, I don't really care how it looks in real life. When I'm doing my make-up, I won't look in the mirror, I'll hold my phone up.'

'So the mirror can fuck off because it's reality?'

'The mirror can fuck off,' she nodded.

As her look for the day came together, she'd snap selfies to record it, adding a little blusher, *snap*, then a little more blusher, *snap*. The process would be repeated with hair and clothes. Next, assuming she was going to her part-time job in the bookshop, she'd get on the bus and continue taking selfies. Sometimes, she'd hear people sniggering. 'I'll just think, "Oh well, they're looking at me and thinking about me and they're probably going to remember me throughout their day."'

'And that's good?' I asked.

'Yeah! Go ahead!'

At work, she'd take selfies in the staff room, and also on the shop floor with customers. She'd be especially delighted when someone else asked to take a selfie with her. 'I'll make a point of having my hands in the shot so you can see I didn't take it,' she said. 'It's like, "Oh, someone wants a picture with me!" I love it.' In the evening, she'd be selfie-ing with the people she lived with in her halls of residence. 'I'll say, "We haven't got a picture today!" and they'll be, "Err, can we skip today?"' Then it's up until 4 a.m., sometimes later, processing and posting. 'I've genuinely taken selfies at a funeral,' she said. 'It was my godmother's. I was in black with red lipstick. We were standing around waiting for the coffin to come in and I was thinking, "This is a look I don't pull off often."' When her mum told her she was being inappropriate, CJ replied, 'I look good. It's always appropriate.'

CJ was slim, pale, elfin and clever, a pretty much straight-A student. Her room at her halls of residence at Roehampton University, where she was studying Drama, Theatre and Performance, was decorated with coloured hearts and glittery stars and photos of Audrey Hepburn. There were posters of vintage Walt Disney films and a motto in a frame – 'Sometimes I feel like giving up but then I remember that I have a lot of motherfuckers to prove wrong.' She had several tattoos, including a large arrow on her forearm. 'It's from *The Hunger Games*,' she told me. 'The main character says, "When I lift my bow up and pull my arrow back there's only one way the arrow can go and that is forward." I've always loved that. Go forward. Don't look back, no matter which direction life seems to be taking you.' She also had tattooed on the skin above her heart the signature of Princess Diana. With her combination of beauty, rebellion and fame, this was the woman CJ most aspired to be.

Her life's ambition was, she told me, 'to be known'. If you're suspecting that CJ might have a few narcissistic tendencies, you'd probably be right. She bravely agreed to take Narcissistic Personality Inventory, or NPI, the scale commonly used by social psychologists to assess the trait. Consistent with her desire to never be less than top in any ranking, the score she sent me back was an impressive thirty-five out of forty. The website on which the test was hosted said this was 97.9 per cent higher than their sample's average.

Although I was never quite convinced CJ wasn't exaggerating to make herself seem the perfect case study (That NPI score! She also said she'd taken 'easily hundreds of thousands' of selfies over the years. Can it be true? She insists it is) she did, in many other ways, seem to be a product of our times. If the ever more gamified individualist economy has left us feeling we have to be increasingly perfect in order to get along and get ahead, then it's easy to see why the self-esteem movement caught on so spectacularly. It told us that there was an easy hack to becoming fitter, happier, better players. Our authentic selves were perfect, it said, and all we needed to do to succeed was to believe it. In the years following the popularization of this idea, much of the data points to a measurable leap in narcissism. And, then, into this culture of self-love came an innovation from the silicon realm: the selfie camera. That taking a picture of yourself and displaying it to the world for comments and likes has become such a lasting phenomenon says a great deal about who we have become. 'People could have done all sorts of things with that technology,' said Professor Campbell, who co-authored Twenge's research into the narcissism 'epidemic'. 'We could've filled the internet with pictures of flowers and architecture and "Mom-ies" – taking pictures of our mom every day and saying how great our mom is. But we didn't do that. Well, we did it a little bit. But selfies are what boomed.' The people who took to selfie-culture with such ease were, of course, the children of the self-esteem generation. That millions of parents spent the eighties, nineties and at least part of the noughties telling their children they were special and wonderful perhaps explains something about the causes of this surge. 'I think so,' Dr Eddie Brummelman, who led the study that found the relationship between parental overpraise and childhood narcissism, told me. 'Whilst we can't conclude that directly from our study, we have seen a rise in these sorts of parenting practices.'

Some of this accords with CJ's stated account of her childhood. Her story begins one weekend when she was seven, when a boy she knew dared a girl she knew to hold a knife up to her throat. They made a little cut. It didn't bother her all that much. 'It was like, "OK, that wasn't very nice of you."' But her parents reacted badly. 'They were completely beside themselves about it.' Then, about a year later, in Basildon Laindon Park, near where they lived in Essex, someone tried to grab her and drag her off. 'It was a bit crap but it didn't actually affect me at the time.' Her devastated parents moved her and her two step-brothers out to a small community on the Dengie Peninsula called Maylandsea. But even out there, on the shores of the muddy Mundun Creek, they didn't find safety for their precious girl. At her new school, CJ was bullied. Her parents couldn't believe it. Yet again, they took action. CJ would be taught within the four brick walls of home. She'd stay inside, loved and treasured, a princess safe in her castle, locked away from the deviants and the cutters and the trolls. 'From then on, my parents just wrapped me in a bubble,' she said. 'They were like, "You're not going anywhere."'

When she was ten, she says, her father's pharmaceutical company started generating fantastic amounts of money. The family moved again, this time to a much bigger house. Her mother gilded CJ's sense of self with compliments. She told her she was beautiful and talented, that she should be a dancer, a writer, a photographer, an actor. She'd leave her handwritten notes that said, 'You have a smile that could change the world. You are going to do amazing things. You are more than you think.' Her father bought her presents. She'd tell her dad, 'Everyone's talking about these Mac computers,' and the next day, there it was: a Mac computer. She'd say, 'I'm bored of my phone; I want a pink one,' and, as if by magic, a pink one would appear. She dreamed of swimming with dolphins in Florida and then, there she was, in perfect blue water, and there, a shining snout. 'I didn't want for anything,' she said. At home, CJ could do pretty much what she liked. Whatever made CJ happy was the rule.

And still the money kept on coming. When she was sixteen they moved again, to a proper mansion, this time, in a town called Billericay. 'It was one of the most lavish houses you could ever imagine,' she said. It had tall security gates, three floors and a vast basement kitchen, with heated marble floors, that was as long and wide as the building itself. There were speakers in every ceiling. CJ colonized one of the spare rooms. She'd play loud music and dance for hours – rap, R&B, contemporary, musical theatre – and her mother would watch her from the back and say, 'CJ, you're incredible. You belong on the stage,' and CJ would say, 'Yes, I know.' They found a private tutor to give her an education, a lovely man who'd been a head teacher. When she wrote an essay about how annoying it was moving house all the time, he told her, 'This is the best work I've ever seen from someone your age.' Her mother was ablaze with admiration. 'You're amazing,' she said. 'You're sublime. You're a prodigy.' And CJ thought, 'Yes. Yes, I am.'

But the mansion behind tall gates in Billericay soon began drawing CJ deeper into its long corridors and freshly decorated rooms. It was her sanctuary, her fortress, the delicate atmosphere that protected her, in all her rare beauty and talent, from the nuisance of other people. She didn't appreciate visitors. When aunts and uncles and cousins would arrive she'd glare at them thinking, 'Why are you in my house?' She'd run to her bedroom, locking the door behind her. One afternoon she was eating in the kitchen when her step-brother walked up behind her. In a friendly gesture, he poked his fingertip into her waist. CJ leapt from her chair and attacked him, pushing and hitting. 'Why did you fucking do that?' CJ hated being touched. She never really felt anything at all, except anger. She couldn't decide whether she was a princess in a castle or a robot in a box. But she couldn't change how terribly she was behaving and, what's more, she realized she didn't want to. Her parents started to worry. 'They'd sit me down and be all like, "CJ, darling . . ." And I'd go, "Darling? Do you want to fucking say that again?" They'd say, "You can't be like that." I'd go, "Why? That's being unfair to me." Then my relatives would chime in and say, "She needs help." My parents would be like, "No, no, she's fine. Let her be happy."'

And then, one day, just like that, CJ started to feel again. Nothing in particular had happened that morning. She'd just woken up feeling so angry she wanted to throw chairs at the wall. 'I'm going to the park for a little bit,' she called to her mum before walking, alone, to her usual spot on the bench. The anger she felt had become measureless, a universe of fury that could no longer be contained inside her narrow body. 'I found a twig and made it super sharp. I broke it on the edge of the bench to crack it in half and I made it thinner and thinner, but just made the top bit pointy by peeling off certain bits. And when I felt it was sharp enough I started taking to my arm.' She drew it, back and forth, back and forth, but it was only giving her friction burns, slim ribbons of pale pink, crumbs of skin. 'What the fuck?' she thought. 'I can't even cut myself.'

She walked home to find the house empty: 'This is perfect.' In her en-suite bathroom, she turned on the shower, in case anyone came back, locked the door and cracked open the casing of a plastic razor to release its blade. It happened, then. 'I just went to town,' she told me. 'Proper slashes.' The pink ribbons were long mouths now. The power filled her. She was immense. Blood was everywhere, on the walls and down her body, spats and drips, vivid scarlet against the white. It looked as if someone had been murdered. Had *she* done that? She crouched down on the floor, suddenly afraid of the person she'd just been. Nevertheless, it was to become addictive. Between the slats of her bed she kept a little keepsake tin she'd got for Christmas. It contained five razor blades (Gillette were the best), some nail scissors and shards of broken CDs. There was also a small bandage, antiseptic cream, lavender oil for burns. CJ only felt alive when she was dancing or cutting. In the pristine instant that the blade entered her skin, she had the control of a god. She could cut how and where she wanted. She could choose to draw blood or not to draw blood. She could see it all happening in microscopic detail, millimetre by millimetre, as if time itself and all the pain in the world were under her command. 'It was like, "I'm powerful, I'm here, I've got a choice to do something and I'm going to do it."'

When CJ announced her desire to attend a school for performing arts, her mum and dad were worried. 'I'm sixteen,' she told them. 'I can do what I want.' And, as usual, CJ got her way. She considered herself the finest actor in her class. She'd pace the studio and the stage, radiating arrogance. When tutors criticized her work she'd tell them, 'That was how I wanted to perform it. You don't understand me.' She'd gladly tell people they were shit. Why not? 'I was being honest, being real,' she said, as the ghost of Fritz Perls grinned in the ether. That proved she was a good person, didn't it? Every now and then she'd catch her classmates looking at her with an expression that said, 'You're such a prick.' But the funny thing was, on the rare occasions she did praise anyone, they'd bloat with gratitude. They began to look to her for validation. 'They were always seeking acceptance from me because I wasn't giving it. I became that person, the coolest one, through just being like, "You're shit."' When one boy demonstrated his exam piece, just for her, she told him he should give up performance and do something academic. 'Why are you here?' she demanded of him. 'Really, why are you here?'

'Because this is my dream,' he said.

CJ replied, with softness and claws, 'Sometimes dreams don't work out.'

It was the greatest performing arts institution in the country: the Brit School. Amy Winehouse had studied there and so had

Adele. The only problem was, CJ wasn't at the Brit School. She'd been sent to this crappy little place in the middle of nowhere. It was a waste of time, for a talent such as hers. After a year, she dropped out. She knew her writing was brilliant and her photography was amazing, so, with her dad's encouragement, she decided to pursue those subjects instead. But at college she found a new distraction. Previously, boys hadn't especially interested her. She had a career to make. Who had the time for all that? But at seventeen, at her new place, she finally realized boys *were* good for something. She put away her hoodies, jeans and Converse and cut her hair into a sexy bob, one strand tucked behind her ear and another falling across her elfin face. And from there on in, it was boys, boys, boys. Boys every night.

The most surprising thing about making men fall in love with you is just how easy it is. It begins, usually, in a bar. You start by making fleeting and inconsistent eye contact. You tangle them into a state of neurotic wondering: is she looking at me? Just when they'd decided they were mistaken, and so felt a moment of embarrassment and vulnerability, she'd throw them a direct, sultry glance, gently biting her lip. And then she'd look at something else for a while. She'd draw this game out for as long as possible, with each play increasing her feeling of control and her target's sensation of torture and shame. Then, she'd throw one right in their eyes and keep it there. With a lingering, intimate smile, she'd finally rescue them from their torment. When they'd summon the courage to approach, 'I'd just be super, super, super, super nice,' she said. She'd move elegantly and slowly with her bum stuck out. 'And they'd come up to me and say, "Oh, you're the most beautiful thing I've ever seen." I'd go, "Oh stop it, you're making me blush." But in my head I'd be going, "Yeah, you're damn right I am."' On a good night, CJ would take home ten phone numbers.

But she wasn't doing this for love or sex. There was a feeling that CJ knew was better than any orgasm: total, teary, puppy-like adoration.

'I constantly needed that validation,' she said. 'And if there was a break where I had no one complimenting me I'd be like, "On to the next one."'

'That was the buzz for you?' I said. 'Someone adoring you?'

'Yeah.'

'And that was about your beauty – being told you were beautiful?'

'Yeah.'

'It wasn't like, "You're a wonderful person."'

'Oh no, I didn't care about that,' she said. 'Angelina Jolie's a wonderful person, but that's not what people pick up on with her. I needed to be seen as being physically beautiful, great, amazing. I needed that validation and I would always get that through the players.'

She'd target the players, because they posed the greatest challenge. If a player fell for you, convincingly and hard, that really meant something. 'I wanted to be in a crowd of girls, and be the one that they picked.' She'd begin pseudo-relationships with them, find out all about them, what they liked and how they thought, and then use the intelligence she'd gathered to conjure herself into their dream woman. One guy she dated adored Eminem. She rapped lyrics at him that she knew from her brothers at home, pretending all the while to be equally obsessed. She even painted a portrait of Eminem and presented it to him as a gift. He was practically speechless: 'This is the best thing ever,' he whimpered. 'No one's ever thought enough of me to do something like this.' But CJ felt nothing for him. 'All I wanted to do was win them so they'd think, "Actually, I don't want to be a player any more. I want to pursue things with this girl." Once I realized that was the case I'd be like, "OK, see you later."'

When I asked CJ to describe the logic of how she saw romantic relationships, she pushed her fingers into the shape of a hill: once you'd reached the point where they unquestioningly adore you, she explained, there was nowhere else to go.

When they truly believed she was better than any girl they'd met, 'I'd be like, "Yeah, damn right I am. Now fuck off."' With the mercilessness of the conqueror, she'd drain the relationship. She'd say she was too busy and reject their calls. They'd text her, begging, 'I really want to make this work, we have such a connection,' and she'd smile, tilt her head and block their number. 'And then you'd get the indirect Facebook statuses,' she laughed. 'They'd post stupid things like, "When you really love a girl but she's not returning your calls", thinking I'd see it and go, "Oh, you love me?" when really I'd be like, "I knew you were going to do that."' Sometimes, they'd try to make her jealous by posting pictures of them drinking with other women. 'I'd go, "OK, I see what you're doing." So I'd send a kind message and make them feel really, really shit. And they'd be like, "You weren't returning my calls, you weren't returning my texts," and I'd go, "I've been going through a really hard time, OK?" And they'd be like, "Oh shit." And in the end I'd go, "You know what, I don't think I'm ready for this, after all, because you really hurt me." And that would be what I did for fun.'

When she was nineteen, something utterly unexpected happened. CJ fell in love. She'd never had sex with any of her players, partly because she hadn't wanted to give away that power, partly because she wanted to be sure they adored her and not her body. But Perry was different. She lost her virginity to him and allowed him to see her bare-faced in a tracksuit watching Pixar films. But then, on his phone, CJ found messages to his ex. 'I looked through the conversations and he was basically telling her everything he'd like to do to her, like push her up against the wall and do her hard, and things like that. Then there were the pictures.'

'Sexual ones?'

'Yeah,' she said. 'The shocker was that he'd take pictures while I was with him. One time he excused himself to go to the toilet while we were out for dinner and he took pictures there.

He actually messaged her and said, "I'm out for dinner with my girlfriend, winky face." I went ballistic.'

The relationship soon ended. She'd expected it to be perfect. She had failed. In its aftermath, whenever CJ went to the fridge for food she'd get flashbacks of the images Perry had sent to his ex. So she stopped opening the fridge. Soon she was going days eating nothing more than a quarter of a digestive biscuit, surviving mostly on Red Bull. The thinner she became, the more she'd stand out. She'd accentuate the effect of her wasting away with contouring and heels. She liked it when people said she looked ill. 'I'm going through a hard time,' she'd tell them, theatrically. She liked it, too, when they said she could be on the catwalk. As long as people were talking about her, CJ was happy. 'Out of all the insults someone could throw at me, like, "You're a bitch, you're horrible and narcissistic," it means nothing,' she said. 'Like, yeah, I'm probably all of those things. But if somebody said, "There's nothing special about you, you're just ordinary" . . . ?'

'That would be bad?'

'Yeah, that would really get me,' she said. 'That would be the worst thing I could think of.'

Today, CJ's selfies, and the comments they generate, are a mechanism by which she feels good about herself. 'Getting a good comment sets me up for the day,' she said. 'I have my friends who say, "You're a beautiful person," or whatever, but for some reason I seek validation from strangers. People online.' She shrugged. 'It feeds me.'

Although seeking approval from social media followers isn't inherently a bad thing, Professor Jesse Fox at Ohio State University's School of Communication told me that the cycle of constantly needing it is. 'What's dangerous about getting accustomed to constant social feedback and constantly being told you're pretty is that when it gets cut off, you start to feel bad about yourself. You need that hit.' Not only does the fact that our mobile phones are always with us exacerbate this problem, so does the structure of social media platforms that place us on the same stream, and afford us the same space, as world-famous celebrities. It invites us to feel a kind of equivalence with them and to compare ourselves to them. What's more, because the new famous are partly comprised of reality TV and social media stars, they seem much more like us than the old-fashioned species. 'For a long time, Hollywood starlets were elevated to the level of untouchable,' Fox told me. 'We always knew these were people who were genetically gifted and more talented than we are. But now anyone can be a star and that makes everyone thirsty.'

All of it adds up to an increasing sense, especially in the young, that we have to maintain a continual state of perfection. 'The problem with selfie-taking and this constant focus on appearance is that you're retraining yourself to think, "What do I look like right now? Do I look fat? Why is that guy looking at my hair? Where's the mirror?" We all know teenagers are horrible to each other. They're in this constant state of heightened awareness about their appearance and about what people might be saying about them.' Fox had a memorable experience of just this kind of thinking at a recent academic conference in Puerto Rico. 'I went to the pool, and I didn't care, you know – I'm a middle-aged woman in a swimsuit, look out! The first people I ran into were my graduate students. They were all sitting there fully covered up by the pool, looking longingly at it. I was like, "Did you all pack swimsuits?" They were like, "Yeah." I said, "Well get in the darn pool!"' Of course, body consciousness in the young is not unusual. 'But it's exacerbated by, "Well, what if someone takes a picture and puts it on social media?"'

Of course, social media is about more than just appearance. It's also a deeply neoliberal product that has gamified the self, turning our identity into a pawn that plays competitively on digital platforms for likes, feedback and friends – the approval of the tribe. The game's winners can ultimately become extremely wealthy celebrities while the losers are often rejected by the group, at times with appalling personal consequences. Most people play away from these extremes, somewhere in the middle of the game, constantly being buffeted back and forth between feeling like a success or a failure. Researchers are not quite able to tell us definitively whether any of this is a cause of depression, although two longitudinal studies have found that more social media use leads to more unhappiness whilst the opposite isn't true (and interestingly, the unhappiness effect only really kicks in after two hours of use per day). 'What we see is studies that show social media is associated with diminished well-being and lower life satisfaction, because people are always looking at other people with better lives,' says Fox. And awareness of the artifice that's inherent in social media is, apparently, no protection. 'In that minute, we accept all the fake stuff as real. We don't question it. And all sorts of bad things can come out of us fixating particularly on our appearance, really nasty effects: eating disorders, diminished cognitive performance, depression, suicidal thoughts.'

As I said goodbye to CJ, on a rainy winter's day in Barnes, I asked about her plans for Christmas. It was late December and we were alone in the kitchen of her halls of residence. The whole place felt abandoned. 'I'm working Christmas Eve and Boxing Day, so I have to stay here,' she said.

'You're not going home for Christmas because of your part-time job at the bookshop?'

'It's one of the catches of it,' she shrugged. 'I know it's the festive season but I've still got a job to do.'

'Bloody hell,' I said.

'I know. When I told my mum she wasn't happy.'

I realized, as I left her, that CJ had told me some things that might seem not entirely pleasant, but I couldn't help but like her all the same. There was something about how unapologetic she was about it all, and how unceasingly willing she was to talk . . . well, to talk about herself. Perhaps it was my culture speaking, but I admired her commitment to authenticity. I was surprised, too, to hear that she was putting her job above the needs of her parents, to whom she remained close. It seemed, in its way, thoroughly neoliberal, as did her devotion to *The Hunger Games*, a touchstone story for her generation having now sold tens of millions of books. Not only was *The Hunger Games* partly inspired by Greek myth, it seemed to me it could almost be read as a parable about neoliberalism, in which a plucky young citizen is forced by the powers-that-be to compete in a brutal public competition, having to fight even those she loves and being told that, in order to live, she must win.

As for the narcissism, I understood, of course, that CJ was an outlier, and that concluding that most millennials resembled her would be absurd. Nevertheless, it felt valuable to put a story to the data that Twenge and Campbell had supplied; to get a visceral sense of the narcissism that had apparently, on the average, been rising. For me, the lack of empathy was the most striking thing – all those poor, confused boys with their broken hearts. There was no map for the world in which they'd found themselves.

Bouts of self-harm aside, CJ, like many millennials, seemed physically healthier than my generation tended to be. When I and my Generation X cohorts were her age, it felt as if we were constantly either drunk or on E or recovering from being drunk or on E. We certainly weren't anywhere near a gym. Indeed, the pressure to be perfect many young people feel today might actually be having some positive effects. Studies suggest millennials are more likely than their elders to avoid risky behaviours including drug use, binge drinking, smoking and teenage sex. They also exercise more and enjoy a better diet.

It's not possible to state categorically that the cultural pressures I've been investigating are the cause of these changes, and even if we did have that data, it's likely those causes would turn out to be mixed. But it doesn't seem unreasonable to speculate that the age of perfectionism, and the neoliberal economy underpinning it, might be at least partly what's driving them. All of these behaviours serve to make them better engines of profit and success and more likely to be winners in the game.

Another striking characteristic of CJ's generation is their greater awareness of the deep structural inequalities that still exist in society. From a neoliberal perspective, the underlying belief that fuels much of this could arguably be that everyone deserves an equal shot at playing the free market and identity should be no barrier to success. Data for the latest cohort, sometimes called Generation Z, suggests narcissism began tailing off sometime after the financial crisis (although some believe this may turn out to be a transition into a more vulnerable, less grandiose form of narcissism).

Other changes emerging in the data feel more unnerving. A study of undergraduates at US universities found students more anxious about growing up than previous generations, and more likely to agree with statements such as, 'I wish that I could return to the security of childhood.' In the UK, another study found four in ten young people describing themselves as 'worn down', with 38 per cent of young women and 29 per cent of young men worried about their mental health. Meanwhile, the principle of free speech seems to be venerated less than ever: a 2015 Pew survey found 40 per cent of US millennials agreeing that the government should be able to prevent speech that's offensive to minorities. For Generation X, that figure was 27 per cent, whilst for the boomers it was 24 per cent.

The rise in individualism that studies have found also seems to have brought with it a dramatic decline in empathy. A meta-analysis by Dr Sara Konrath at the University of Michigan indicated that today's college students are 40 per cent less empathetic than students in the 1980s. They're less likely to agree with statements such as, 'I often have tender, concerned feelings for people less fortunate than me,' and 'I sometimes try to understand my friends better by imagining how things look from their perspective.' They also trust less.

Pew Research data suggests only 19 per cent of millennials believe most people can be trusted, compared with 40 per cent of boomers. Although bullying at school is down, bullying online is rising.

It's on the internet, in particular, that the beliefs of others are policed, their heresies loudly punished. It occurred to me that this might be the next irresistible step on the road we've taken: if we are all gods, then our feelings are sacred, and if our feelings are sacred, the people who hurt them must be sinners. It felt as if, at the heart of all this, there was something inherently narcissistic – that our perspective is so precious we feel justified in silencing or punishing those who don't share it.

I wondered if this was driving the increasing intolerance that's been reported on college campuses, which have seen visiting speakers banned and campaigns to have professors punished or fired. Recent years have seen the rise of the aggressive and increasingly violent Alt Left. At Middlebury College, Vermont, one professor required hospital treatment after being assaulted whilst escorting a controversial speaker to a talk. At Evergreen State College, Washington, the liberal biologist Professor Bret Weinstein was surrounded by Alt Left students and threatened with violence after complaining at proposals to discourage white people from attending college, for a 'Day of Absence', on the basis of their skin colour. When students at Matteo Ricci College in Seattle complained that the focus on Western classics in their humanities course was 'psychologically abusive', they demanded the resignation of the Dean, Jodi Kelly. She was placed on administrative leave and soon retired, to the delight of the protesters, who called it 'a success of years of organizing'. A comical and well-reported case featured a push to have the vendor of inadequate sushi in a college dining hall expelled for sins including cultural insensitivity and appropriation ('If people not from that heritage take food, modify it and serve it as "authentic," it is appropriative,' wrote one protester). In the UK, University College London banned the

Nietzsche Society because its existence was a threat to the 'safety' of students.

One remarkable feature of this new discord is the language students are using to denote injury. They speak of challenges to their points of view as acts of 'violence' or 'abuse' which leave them 'unsafe' and 'traumatized'. It's as if their inner self, their 'soul', is so precious as to be sacrosanct. Also notable is the me-focused direction of much of their political activity. Whereas older generations protested in empathy with distant peoples – in apartheid South Africa, in Vietnam, in Biafra and in solidarity with those in Latin America or the Caribbean who were suffering as a result of US foreign policy – today's privileged, angry students seem far more preoccupied with changing the world for themselves and those near them.

When I asked Professor Campbell for his view on whether the self-esteem movement and the 'narcissism epidemic' he'd investigated had any role in this, he said, 'For me, as a professor at one of these colleges, let me just put a very clear "no comment" on this.'

'It's too dangerous, is it?' I asked.

'It's very dangerous, yeah.'

BOOK SEVEN

How to Stay Alive in the Age of Perfectionism

Austen Heinz stood at the window of his apartment, the street lights of San Francisco reduced to dots of dancing fire in the movement of his eyes. Dressed in his usual outfit of flip-flops, blue jeans and a North Face fleece over an unbuttoned shirt, his rock-star-ish hair fell across his shoulders. At school, back in North Carolina, he'd been bullied, sometimes badly – the combination of his physical slightness and social illiteracy had seen to that. They'd called him 'the professor'. He had a kind of genius for absorbing large amounts of complex information at great speed and when he arrived at an opinion, often after a period of intense labour, he'd announce it provocatively and unapologetically. Austen was bad at reading people. He'd struggled to find friends and with his mental health. But now, at the age of thirty, he was living a few blocks from his own laboratory, in a high-rise apartment in San Francisco's tech district, looking out over 180-degree views of the Bay Bridge, the AT&T ballpark and the stunning harbour. He was a Silicon Valley founder, creating prosperity and progress, advancing the human cause with genius and toil: an Ayn Rand hero come alive. His idea and his company, Cambrian Genomics, would change everything, of that he felt sure. As he'd tell investors and journalists again and again, the tools they were developing would one day be more powerful than the hydrogen bomb.

Imagine that it was possible to create the perfect human. The process would be like making an app, but instead of computer code, your design language would be DNA. You'd do the creating itself on your phone – using a piece of software called a Genome Compiler – then email what you'd come up with to Austen's laboratory. They would manufacture the DNA, as per your instructions, then send it on to you. They'd dry it out and pop it in the post. After all, DNA isn't alive. It's a polymer, an arrangement of four different chemicals. From that DNA, it would be theoretically possible to construct the most advanced forms of life. You could make your human a super-genius, immune to all kinds of diseases. You could make them live for ever. After all, we only age and die because the DNA program we're running – our human code – contains an instruction to do so. Just get rid of it. Rewrite it. Why not?

And why stop at humans? You could design and create any form of life you'd like. In the future we wouldn't leave it to messy nature just to plop everything out, riven, as it always is, with all those hundreds of thousands of little genetic imperfections that add up to sadness, illness and death. Everything would be synthetic, designed for purpose, including our children, including us. The only things that would limit us would be our DNA programming abilities and our imaginations. Eventually, we wouldn't even need to use Cambrian Genomics' expensive equipment. We could construct a *Chickenasaurus rex* on our screen and, minutes later, see it scuttling from a printer. And there wasn't even any particular reason why we'd be stuck remixing the things we already had. Everything alive is made up of just twenty different amino acids. Why not expand the range? Make some new ones? Incorporate metals, say, into plants or animals? Imagine the possibilities. Imagine the problems we could solve.

If all of what Austen was planning was to happen, it wouldn't only give us *Jurassic Park*, it'd give us *Terminator* too. But it wouldn't necessarily follow that the Cambrian Genomics future would be a disaster movie. By curing all disease, living for ever and solving some of the planet's most enduring technical problems without destroying it in the process, we could reduce the sum of human misery considerably. It's true, of course, that only the lunatic talks earnestly of paradise. But how crazy do you have to be to think that, with this technology, we could move ourselves an inch, even a mile, towards it? If Doug Engelbart was the visionary of the Information Age, then perhaps history will one day crown Austen one of his heirs: a restless, optimistic, socially-maladjusted prophet of the oncoming Synthetic Age in which the project isn't to augment human intelligence, but humans themselves.

It was at Duke University, whilst working on a synthetic biology research project, that Austen came up with a new and efficient way of producing usable DNA that reduced the cost from tens of thousands of dollars to just a few. 'Everyone else that makes DNA, makes DNA incorrectly and then tries to fix it,' he said. 'We don't fix it. We just see what's good, what's bad and then we use the correct pieces.' This drastic reduction in cost would enable them to treat DNA like we treat data – as cheap to make and emailable, programmable. When he was in his mid-twenties, at Seoul National University in South Korea, Austen developed his concept of a 'printer for DNA' that would do the selecting, then decided progress would be more rapid in the private sector. At twenty-seven, he returned to the United States with some bridges burned and $300 to fund his vision. He was just another west-coast brain convinced he was going to change the world.

As a kind of trial run for his technology, he and some colleagues decided to create a glowing shrub by copying some DNA code from a firefly, printing it out, and inserting it into the cells of a plant. Their test worked. Their hybrid plant glowed in the dark. They decided to put it on sale. The $10,000 they spent on a promotional video was quickly recouped; they took in almost $60,000 on their first day of sales and $484,013 within six weeks, with orders eventually building towards $1,000,000. He founded Cambrian Genomics and raised $10,000,000 from venture capitalists including PayPal billionaire Peter Thiel. His company began partnering with major international corporations, such as Roche and GlaxoSmithKline, as well as some smaller start-ups. One of these was

Sweet Peach, which had been founded by Audrey Hutchinson, a young biology student and Distinguished Scientist scholarship recipient at New York's Bard College. After suffering a series of painful urinary tract infections, Hutchinson had become interested in vaginal health. Hearing about his work, she emailed Austen with her idea for a company that would use Cambrian Genomics tech to manufacture vaginal probiotics. Customers would send in a swab that would be genetically sequenced. Once the specific microbial species that made up their particular bacterial community was analysed, a personalized treatment would be delivered. Austen was immediately interested. He agreed to help not only with the technology but with business advice. He took a 10 per cent stake in her company.

Word of his work spread further. He met Sergey Brin from Google, Elon Musk from Tesla and SpaceX and Jared Leto from the movies. He was invited to Richard Branson's private island, where apparently he silenced the billionaire's dinner table with his visions of an intentionally designed, synthetic future. He was interviewed by *Fortune* and NPR and *Wired*. CNN named his technology as one of its 'Top Ten Ideas That Could Save Lives'. He became a frequent guest at tech conferences, one of which was Demo: New Tech Solving Big Problems. His presentation, in San Jose on Wednesday 19 November 2014, was entitled 'Create Your Own Creatures by Printing DNA'. 'Our goal is to take everything that's existing and natural and replace it with a synthetic version,' he said, with his familiar outfit of blue jeans, shirt open at the neck and fleece in place. 'So by writing DNA we can make it better. We can make better humans, we can make better plants, we can make better animals, we can make better bacteria.' His glowing plant project might sound trivial, he acknowledged, but the implications were immense. 'If you can engineer a plant to glow in the dark, imagine what else you can make a plant do.

You could make a plant suck all the carbon out of the atmosphere. You could make a plant that produces food to feed the world.'

The day before the conference, Austen had apparently been told he would be on for ten minutes rather than the three he'd been planning. To fill some of the time at the end, he decided to speak briefly about some of the companies he'd partnered with who'd be using Cambrian Genomics technology. Welcoming one of these partners onstage, Gilad Gome of Petomics, he talked about the idea of changing the smell of faeces and gastric wind and using it as an alert that a person was unwell. 'When your farts change from wintergreen to banana maybe that means you have an infection in your gut,' he said. He introduced Sweet Peach as a similar project. 'The idea is to get rid of UTIs and yeast infections and change the smell of the vagina through probiotics,' he said. 'So not only can you actually program them, you can write them, you can change them and you can make them personal to you. You can control all the code that lives on you, which is exciting, because previously the natural world has been beyond our grasp. We've recently, within the last ten years, been able to read it. Now we finally have the cost low enough that anyone can write it on their phones. So the idea is, your microbes can be out of balance. Sweet Peach will balance them, improve smell, and everybody's happy.' Everybody's scents are bacterial in origin, he explained. They're produced by organisms that live on you. 'We think it's a fundamental human right to not only know your code and the code of the things that live on you but also to write your own code and personalize it.' When the compere provocatively asked if Austen and Gome were playing God, Austen countered in exquisitely neoliberal fashion, 'The idea is personal empowerment. We don't want the state telling people what they can grow on them, what babies they have and what genes they can fiddle with. We want it to be self-directed.'

Down in the audience, a journalist from Inc.com thought

what he was hearing was 'astonishingly sexist'. After all, here was a man, he'd later write, chattering about 'making women's sex organs more aesthetically pleasing'. It seemed to him that Austen was just another of these 'tech bros' who 'talk endlessly about changing the world with technology while building frivolous things'. After the presentation, he asked some follow-up questions. Gome explained to the reporter that the change in scent wasn't only there to help customers connect to themselves in a 'better way', it was an indicator that the product was actually working. 'It tells us where the protein is expressed,' he said, adding jokingly, 'What, would you rather have it glow?'

'These Startup Dudes Want to Make Women's Private Parts Smell Like Ripe Fruit' ran the headline at Inc.com later that day. The story zipped around the web, being swapped and swapped and swapped again on social media, the outrage rapidly amplifying. Soon, the *Huffington Post* picked it up: 'Two science startup dudes introduced a new product idea this week: a probiotic supplement that will make women's vaginas smell like peaches.' Gawker called it a 'waste of science' and said Sweet Peach 'sounds like a C-list rom-com with a similarly retrograde view on the priorities of the contemporary human female'. Then, Inc.com weighed in again: 'Its mission, apparently hatched by a couple of 11-year-old boys still in the "ew, girl cooties" stage, is to make sure women's vaginas smell "pleasant".' Similarly negative stories began appearing in major news sources such as Salon, BuzzFeed, the *Daily Mail* and Business Insider.

These reports were profoundly unfair. Austen and Gome were presented as misogynists who'd decided to concentrate their efforts on solving the problem of disgusting vaginas. In truth, Austen had spent the majority of his talk explaining the fantastic world-changing possibilities of his technology. Its title referenced not vaginas but creating 'your own creatures'. He'd talked about plants that could counter the effects of global warming, of one day being able to feed the world. Then, as a postscript, he'd mentioned other applications they were developing with third parties. Austen had mentioned 'vaginal smell' in a way that wasn't entirely clear, but was in the context of a discussion of health products. After the presentation, Gome had actually clarified that part of the reason for the smell change was to show that the product was working. It's relevant to note that most of the news sources now attacking Austen and Gome were free to the online reader and so required large audience sizes to monetize. They were quintessential products of the internet age, relying for much of their survival on the sowing and harvesting of moral outrage.

It was Inc.com that stoked that outrage yet further with a follow-up interview with Sweet Peach's Audrey Hutchinson. 'Sweet Peach Founder Speaks: Those Startup Dudes Were Wrong About My Company' ran the headline. 'When I wrote earlier this week about a new probiotic supplement called Sweet Peach engineered to make women's vaginas smell like fruit, the response across the internet was understandable outrage: Who the hell were the guys behind this and what right did they have to decide how women's bodies ought to smell?' But the real story, wrote Inc.com, was 'outrageous in a different way'. It wasn't only their failure to acknowledge Hutchinson. Inc.com accused the founders of being 'highly misleading' by characterizing Sweet Peach as a tool for making vaginas smell like fruit – which, of course, they hadn't. 'It was Gome who introduced the critical misperception about Sweet Peach,' he said, 'after I specifically asked him whether the supplement was designed simply to eliminate unwanted odors, or whether it was meant to introduce desirable new ones, like the scent of peach. He insisted it was the latter, likening the new scent to a marker dye that let the user know the product was working. "Instead of color, this is a scent or a flavor. But it's way cool that it smells good," he said.' Hutchinson told the reporter she'd been nauseated by what had happened, and even claimed to have vomited twice. 'A vagina should smell like a vagina,' she told the *Huffington Post*, 'and anyone who doesn't think that doesn't deserve to be near one.'

Austen tried to rescue the situation. He apologized for leaving Hutchinson out of his presentation, explaining that he was only informed that his three-minute talk had been extended the day before and had lacked sufficient time to plan. Gome, he said, had spoken about Sweet Peach because he was excited about the science. 'He's a microbiologist and he likes to talk about possibilities.' In his typically socially-deaf way, he added that whilst the publicity was losing him investors, it would be good for Hutchinson. 'This mischaracterization is going to be great for Sweet Peach.' He also desperately explained to the *Huffington Post*, 'I never said anything about making vaginas smell like peaches.' None of it made any difference. On 24 November, Hutchinson released a series of tweets in support of Austen. 'Amidst chaos, I'm confident in saying I'm still proud to have Cambrian Genomics as a stakeholder in Sweet Peach,' she said. 'Austen Heinz of Cambrian Genomics has shown me nothing but fervent support in my efforts to make Sweet Peach a force in women's health. He's been a friend and support throughout this entire process and has played a huge role for helping make my vision and company a reality.' But this statement, by the young female founder, was largely ignored.

And the monstering continued. 'How Two "Startup Bros" Twisted and Took Credit for a Young Woman's Company' ran a headline on the *Huffington Post*: 'Yup, you read that right; these "startup bros" think a vagina that doesn't smell like a peach is a Big Problem to be solved.' The *Daily Mail* posted another story ('Female CEO of vaginal probiotic is "appalled" by male colleagues who misrepresented her product to the public'), as did BuzzFeed ('the two completely mischaracterized the company . . . it does not create a peach scent for women's vaginas'). The *Guardian* ran four negative stories over the course of just three days. Slate asked, 'Who is giving these men the time of day?' whilst the Daily Dot wanted to know, 'Is the Sweet Peach startup a complete scam?'

'It was pretty heart-wrenching to see him suffer like that in the media,' Austen's sister, Adrienne, told me. We were talking in the central San Francisco consulting room where she works as a clinical psychologist, her client base largely Silicon Valley tech workers. 'It was click-baity stuff. Article after article after article got written because the headline was interesting. It was so infuriating. I don't think I realized how devastating it was for Austen until later. He couldn't stop talking about it.'

Behind the scenes, Austen had been trying to convince Hutchinson to include a smell signal in her product, but she'd resisted. She'd had no idea he was planning on talking about Sweet Peach, even as a relatively brief aside following his main talk. That he didn't think to mention her name only added to the problems. But, said Adrienne, his presentation contained no malice. She described him as 'one of the biggest feminists I know. I mean, he grew up with two sisters who owned him.' In his attempts to repair the situation, he'd only succeeded in making things worse. 'He was just saying all the wrong things,' said Adrienne. 'I mean, you could never describe him as socially graceful. The reporter was a really nice person but he got Austen completely wrong. He thought he was just kind of a dirtbag.'

Because of what was going on in the media, investors began backing out of Cambrian Genomics. One of Austen's business advisers compared his reputation, in the industry, to that of Bill Cosby. He'd been trying to raise a second round of funding and now he thought he'd have to start laying people off. The timing was terrible; they'd been encountering difficulties with the laser and needed all the brains they could get. 'The technical problems could've been addressed,' said Adrienne. 'He had this brilliant team of scientists that were helping and, worse-case scenario, they could've sold to another company who could've figured it out, or they could've persevered and eventually figured it out. That wasn't the issue. It was more his confidence in his ability to raise money after this media fallout.'

By the end of 2014, Austen was suffering physically. 'He was like a walking corpse. He'd stopped eating, stopped sleeping. He was just so ruminative – there was a constant stock-market ticker of how his life was over.' They'd have long conversations on the phone. 'You might feel like you'd got somewhere by the end of the conversation but then a couple of days later he was back to the same headspace.'

In March, Austen ordered a selection of ropes from the internet and tried to hang himself in his apartment. He failed. When he came too, he called Adrienne, who was driving home from work. 'I just tried to kill myself,' he told her. The family took him on a break to wine country and staged an intervention, one evening, after dinner. 'He just kept saying, "I'm dead, I'm dead, I'm dead." We said, "We're going to take you to the hospital. We're going to get you the help you need."' They had him committed in San Diego.

On 27 May 2015 a member of the Cambrian Genomics team opened up the laboratory, after the long weekend, and discovered his body. Austen had hanged himself. He was thirty-one.

Shortly before his death, Austen had stayed with his best friend, Mike Alfred. 'He felt like the whole world was against him,' Alfred told me. 'He took it a lot more personally than I'd advise someone to.'

'But it was personal,' I said. 'They were calling him a "tech bro" and a sexist.'

'Yes, it was personal. And he took it like that. He took it pretty hard. He talked about killing himself every day.'

I asked if there might have been any truth to what they were saying. There was some evidence of sexism in Austen's past. In 2009 he'd self-published a semi-fictional memoir that contained some unpleasant and juvenile talk of strippers and orgies. 'It was not true at all,' he said. 'He had strong opinions about whether people were smart or not. He didn't have a lot of respect for people that were dumb. But it wasn't gender. He definitely was not a sexist.' The problem, said Alfred, was a lack of social sensitivity. 'He wasn't a person who sat around saying, "How can I make sure that what I'm about to say to this person comes across right?" He would just say it.'

'But isn't that a common personality type in tech?' I asked.

'I think so. There are a lot of really talented people that are really bad at reading others.'

Alfred also described him as a 'tormented soul'. Depression had long been a problem for Austen. Following his death, it was said he'd suffered from bipolar disorder, a claim that seems at least partly based on what he'd written in his semi-fictional memoir. Adrienne disputes this. 'He never received a formal diagnosis of bipolar disorder,' she said, citing his medical records. 'I was bothered that this was published and not fact-checked. The only formal diagnosis he received was major depressive disorder. Austen saw mental-health providers at various points in his early adult life. He was not mentally ill or depressed his whole life. It came in waves and his depression was usually triggered by a difficult and stressful life event.' Ultimately, she said, it was the media assault that tipped him into his final decline. 'No one's to blame for his death,' she said. 'But make no mistake, I know for a fact that this is what initiated this depression episode.'

When I asked Adrienne if she'd describe her brother as a perfectionist, she nodded. 'He struggled with the black and white thinking that can be part of that; catastrophizing – "I'm going to be homeless, everybody will think I'm a failure" – mind-reading of what other people think. And so when he started running into difficulties, he just got stuck in some really severe thinking traps.'

Austen was, in many respects, a victim of the age of perfectionism. If he was the type of person who was more sensitive to signals of failure in his environment, then the environment in which he found himself was savage. Despite his achievements, despite his incredible vision, despite the possibility that his work might truly have changed the world, Austen was the wrong person, at the wrong time, with the wrong personality. He was not always a charming presence and could certainly come off as arrogant and dismissive. But he didn't deserve the treatment he received from the media. There was no justice in the mobbing he endured and neither was there mercy.

Two of the academics I've met on my journey have mentioned this especially unpleasant aspect of our times. 'It's something that's becoming more salient,' Professor Gordon Flett, the expert in the dangers of perfectionism, had told me. 'When a public figure makes a mistake there seems to be a much stronger, more intense and quicker backlash. So kids growing up now see what happens to people who make a mistake and they're very fearful of it.' The other was Professor Kip Williams, the social pain specialist. 'You see it on both sides, from the right and left,' he'd said. 'There are strong pressures to conform and an immediate response to disrupt or to ostracize people who disagree.' The irony of the new digital world that's enabled the rise of these kinds of incidents is that it relies for its success on some of our most ancient parts. We're tribal, and we're wired to want to punish, sometimes savagely, those who transgress the codes of our in-group. These are powerful and dangerous instincts that can easily overwhelm us. For Austen, they triggered a sequence of events that would be fatal.

What Austen's story suggests is that the self is a trade-off. Mike Alfred told me there are plenty of talented people in tech who, like his friend, 'are really bad at reading other people'. They're good at one thing, but that seems to leave a deficit somewhere else. They're not like those heroes at the climax of our stories, that perfect blend of strength, power, empathy and caring. This rule goes for all of us. No matter how hard we try, there are some things we're not that good at, some ways of being we just can't master. Regardless of all the promises we might make to ourselves and our loved ones, there are personal qualities we'd love to have, but can't make stick.

One of the dictums that defines our culture is that we can be anything we want to be – to win the neoliberal game we just have to dream, to put our minds to it, to want it badly enough. This message leaks out to us from seemingly everywhere in our environment: at the cinema, in heart-warming and inspiring stories we read in the news and social media, in advertising, in self-help books, in the classroom, on television. We internalize it, incorporating it into our sense of self. But it's not true. It is, in fact, the dark lie at the heart of the age of perfectionism. It's the cause, I believe, of an incalculable quotient of misery. Here's the truth that no million-selling self-help book, famous motivational speaker, happiness guru or blockbusting Hollywood screenwriter seems to want you to know. You're limited. Imperfect. And there's nothing you can do about it.

*

It was graduation day and a student eagerly approached the Professor of Psychology Brian Little, her parents following just behind. 'Oh, Mom! Dad! I'd like you to meet Professor Little,' she said. 'He was my favourite. He taught me the one lesson that really changed my life. He said, "No matter what, there's nothing you can't accomplish."' Recalling this moment, Brian told me, 'Two possible responses coursed through my mind. One was what I said, which was, "Well, isn't that interesting?" The other was, "You're full of shit. You obviously didn't listen to a thing I told you."'

Brian, who teaches personality psychology at the University of Cambridge and at Harvard, partly blames cultural leaders such as John Vasconcellos for this belief in the all-powerful self, which today seems so widespread. 'The self-esteem movement has had some very negative effects,' he told me. 'The lecture that I give on it is one in which normally ebullient students become really pensive and maybe even saddened. They're very disillusioned when I tell them about the myth of unlimited control.' One particular student he encountered dreamed of becoming a gynaecologist. 'And I really wanted to scream at her, "For the sake of the wombs of the world, do not go into gynaecology!"'

Personality psychologists such as Brian are perhaps best known for their work on the basic human traits, usually defined as openness, conscientiousness, extraversion, agreeableness and neuroticism. The idea is that a human personality can be roughly reduced to a combination of these five qualities. To find out mine, I took a test called the Newcastle Personality Assessor. I was pleased to discover I was high in openness (which is about curiosity, art and adventure), unsurprised to discover I was low in extraversion and agreeableness (grumpy loner) and curious to note I was also high in neuroticism. To understand more about that troublesome-sounding category I found a book, *Personality*, by Professor Daniel Nettle, one of the test's designers. What I discovered in its pages left me open-mouthed, pacing my study, muttering obscenities. I read, for example, that it's 'beyond doubt that neuroticism causes awful, private, lifelong pain, hidden away behind curtains and doors, for millions of people'. It's not merely a risk factor for depression, 'it is so closely associated with it that it is hard to see them as completely distinct'. Its cruellest feature, it said, was that 'the worried actually have more to worry about than the carefree. Study after study has shown that not only do high scorers react more strongly to negative life events, they have more negative life events to react to.'

But the most startling thing of all about this body of research is that people can take the test at several points in their lives and their personalities don't tend to dramatically change. They pretty much are who they are for ever. 'Seductive though it might be,' writes Nettle, 'there's no more point in me wishing myself to have a higher or lower level of extraversion than there is to wish I were, say, born in 1777.'

I made an appointment to meet him. We spoke over lunch at a noisy student cafe in the centre of Newcastle. He was pale, thin, with a runner's face, eyes bright, cheeks slightly sunken. On our way to the restaurant I expressed my surprise that these ideas weren't more widely known. He'd said something about personality psychology lacking 'prestige'. As he ordered his prawn salad, I asked him what he'd meant. 'Is there not yet enough data or something?'

'There's endless data, but it's really crap,' he said. 'It all comes from questionnaires. So people think, "Come on, you're just getting one question that asks something in one way and another that asks the same thing in a slightly different way and saying those two things are correlated." But I think that's changing. People are beginning to say, "What are the brain differences that go with these different responses?" Or, in one or two cases, "What are the genetic differences?"'

Essentially, the hypothesis has it that, just as we're born with differences in hair, hips and voices, we're also born with differences in our brains. The genes we inherit are important in determining how they're wired up. They also regulate brain chemistry (our serotonin system, for example, is mostly in place at birth), which influences not only how we perceive the world but how we react to it. These variations in brain circuitry should, if the theory holds, map onto the five different personality traits. 'Take extraversion,' he said. 'There's a huge literature on brain mechanisms of reward, pleasure and acquisition of stuff. Those mechanisms relate pretty well to extraversion.' Similarly there's a long literature on the different ways serotonin and a brain region called the amygdala can affect anxiety. 'That's panning out as relating to neuroticism.' The cause of variations in traits such as openness and agreeableness remain largely mysterious. It is well known, however, that damage to the brain in specific places can cause specific personality changes. Everything from dementia to tumours to physical injuries have been shown to turn people into killers, thieves and paedophiles. 'A lot of these people were previously model citizens and started doing stuff on impulse,' said Daniel. 'That's conscientiousness, which is a nice example because, if these five traits are true, you should be able to find mutations or brain injuries that affect one but not the other four.'

Although, as we've seen, our neural wiring is by no means complete when we're pulled from the womb; the core of who we are is set at the early stages of life during which we're effectively helpless. We're part nature, part nurture, formed by biology, culture and experience. That non-biological part might make it sound as if we're free after all. But, biology aside, the type of person we become is principally defined by childhood events we have no control over and have very little capacity to reverse. By the time we're old enough to really understand what our personality is, and begin wondering if there's anything we can do to change it, most of the work has been done.

None of this appears to tell us anything happy about our ability to do whatever we want to do, or be whoever we want to be. I thought of John Pridmore who, even in his Catholic phase, was still recognizably John Pridmore when he was thumping the man who spat on his car. I thought of Vasco and his enduring battles with anger and depression, which continued despite his claim to have drunk the magic potion of self-esteem. I thought, too, of my experience at Esalen. Being so warmly and generously accepted by the group had felt transformative, at the time, but the truth of the matter spilled out of my mouth when, without thinking, I'd grumpily muttered, 'Fucking idiots,' at the hippies. Then there was that devastating speech at the Institute's Spiritual Tyranny conference: 'The best therapist turns out to have a clay heart. And Fritz was a dirty old man. And Freud couldn't give up cigars. And Bill Schutz doesn't jump for joy.' And who could forget the words of the former lover of Esalen's co-founder, Michael Murphy: 'He got a million dollars' worth of advice from some of the best psychologists in the country. None of it helped.' It seems a constant, almost an archetype: the guru who claims to have discovered the answer, yet manifestly fails to change.

Which is not to say, of course, that we can't or don't change at all. People grow up and mature, they learn things, become wiser and more expert, traumatic events can damage them and markers for neuroticism fall when they recover from periods of personal crisis. It's thought that therapeutic interventions can affect (but not transform) traits. There are also various predictable shifts in personality as we get older – we often decrease in openness as we exit middle age, for example – just as changes take place in response to environmental pressures: cultural shifts can trigger rises in narcissism, in certain kinds of people, and a state of war can trigger general rises in anxiety. 'The idea of stability of personality shouldn't be taken as meaning that we can't change anything at the average level,' said Daniel. 'People in the Gaza Strip are super-anxious. But even within the Gaza Strip some people are more anxious than others.' But what this doesn't mean is that an individual can, by sheer force of will, transform their fundamental nature. Although genes are not destiny, they do set certain limits. I told Daniel that, having read his book, I'd come to think of adult personality, not as a straitjacket, but as a prison cell inside which we can move towards the window a little. 'Your metaphor might not be quite perfect,' he said. 'You can make a real effort to, say, improve your social relationships or stop being a workaholic and you might manage that for a while. But habits will out. It's almost like there's a drift back in.'

'So a prison cell with a sloped floor?'

He shrugged, perhaps.

It's important to note that these five traits are not like switches. We're not one thing or another. Rather, they're dials that are set higher or lower in each of us. But those dials' fine tunings, which are set by tiny changes across thousands of genes, can have dramatic effects on our lives. Take agreeableness. Imagine how many interactions you have every week in which it could possibly be imagined you've been rejected in some way. These 'rejections' could range from overt acts of aggression to the subtlest interpretations of tone or body language. Your trait settings will not only guide how many of these incidents you're actually aware of, but how you react to them. Some people will tend to shrug and think the best whilst others will become angry, paranoid and vengeful. Others still won't even notice all but the most obvious confrontations. It's another alarm system. We all have slightly different thresholds at which our bells start ringing, just as we have different responses to their sound. Because of a small difference in your amygdala, you might react angrily to one more perceived sleight in every ten compared to a person who's marginally more agreeable than you. But that one extra conflict per week, say, will be enough to give you a very different experience of life in general. You'll likely have a different reputation, at work and amongst friends and family. You might have different implicit beliefs about the nature of people, of authority, of your social sphere. You'll probably think differently about yourself. That small distinction, based on the way your brain happens to be wired up, is enough to gather you up and dump you into another life.

Through the processes of 'social evocation' and 'social selection' these differences can create an environment for us to live in that only serves to reinforce our natural tendencies. Our personalities, to a significant extent, make our worlds. We tend to socialize with people who are similar to us, just as we gravitate towards communities and forms of employment that suit who we are. There can, perhaps, be no better example of this than the monks of Pluscarden, not least Father Martin, who appeared to so desperately need an extreme state of quiet and routine – and who seemed so wonderfully happy in the one he'd found. Our personalities also trigger responses in others that echo our temperament: grumpiness evokes grumpiness, cheeriness invites cheer. As highly social animals, the worlds we live in are largely ones of relationships and interactions. Our personalities help define the texture and temperature of our social worlds. The experience of everyday life is often completely different for people who are living side by side.

These social responses, in turn, modify the things that happen to us. Extraverts really do talk more and get invited to a greater number of parties, so a self-reinforcing pattern begins. Personality is like a prevailing wind that pushes the little daily events that make up a life in a certain direction. It also predicts a wide range of dramatic future outcomes: high neuroticism is a strong predictor of divorce; low conscientiousness is thought to increase risk of death, in any given year, by around 30 per cent; extraversion is a predictor of promiscuity. Personality even makes itself felt in our appearance and surroundings. Studies indicate that extraverts tend to swing their arms more when walking, that people high in openness often look scruffy and that neurotics are more likely to wear dark clothes and have posters with motivational quotes displayed in their offices.

The particular mechanics of your brain write your biography in other surprising ways too. A major study that tracked the lives of six hundred US men found unhappy childhoods *not* to be a significant factor in alcoholism. Other work suggests that the people who are more likely to become addicted are, in fact, the ones who react to alcohol differently: they're the euphoric drunks who, for reasons of biology, happen to be less sensitive to its sedative effects. Those who feel sleepy after a drink are less likely to develop a problem. A person's genotype also changes how they're affected by violence and childhood abuse, just as it alters the likelihood that they'll become winners in the neoliberal game. Individuals who experience greater cognitive reward from winning are more likely to become members of the envied 1 per cent. 'Not everyone who's competitive will be a millionaire,' said Daniel, 'but everyone who's a millionaire will be competitive.' Conscientiousness, too, is positively correlated with prosperity. Much of how successful we are in meeting our model of perfection greatly depends on neural mechanisms that already exist within us and that we have little capacity to change.

Another personality finding that runs utterly counter to the shibboleths of our culture is that, genetic inheritance aside, parents have no systematic influence on the personalities of their children. We know this, Daniel told me, partly from studies of identical twins who are separated at birth and nevertheless end up temperamentally matched. 'And actually,' he added, 'siblings raised in the same family are surprisingly different from each other too.' Which is not to say that parents have *no* effect, as the study into parental overpraise and narcissism demonstrates. It's just that environmental factors, including parental style, trigger different effects in different children, depending on how they're wired up. 'You take two kids and put them through austerity and they might respond in different ways,' he said. 'One might become incredibly driven, spurred on by this early adversity, and the other might be damaged.' These reactions are so inconsistent and unpredictable that parents tend to radically overestimate how much control they have over how their children turn out. 'The main reason to be nice to your children is the same reason to be nice to the person who runs your local shop,' he said. 'You're going to have to see them a lot.'

'So there's no point being nice and hoping they'll turn out nice?'

'They might turn out to be really horrible anyway. Your children will probably be screwed up whatever happens, so you don't need to worry about whether it's you that did it.'

This perspective casts both light and nuance on the reported rise in narcissism. What Jean Twenge and her colleagues detected, Daniel believes, is a shift in the population average with some personality types – namely those who are both high in extraversion and low in agreeableness – being far more susceptible to the rise in individualism and the effects of the self-esteem movement than others. It's these people that would've been more vulnerable, in this specific way, to environmental effects such as overpraise, during their earlier years when their identity was still very much forming.

Some sceptics of these ideas point to the work that shows we can become different people in different contexts: we're one version of us with our mother, another with our lover, and so on. But it's important to remember that these are still *versions of us*. Behaviour is a result of a combination of situation *and* genes. The social world is a maze of circus mirrors, each exaggerating one facet of our self and diminishing another, while the essential core remains. The average level of trait expression might shift from situation to situation, but who you are compared to everyone else in that situation will usually remain roughly stable. A high neurotic on a rollercoaster, for example, will still be among the most anxious people on the ride. But we also have the ability to exercise what are known as 'free traits'. An introvert can obviously adopt extravert traits, in order to make a public presentation, say, or appear gregarious in a job interview. But when we act 'out of character' we can feel it. Writes Little, 'a biogenically agreeable woman who is required by her law firm to suppress her pleasantness and act aggressively may experience signs of autonomic arousal – such as increased heart rate, sweating, muscle tension and a stronger startle response'. All of this contradicts what I was led to believe when investigating the so-called 'self illusion': it turns out that there *is* such a thing as an authentic self, after all. It might not be God, as the intronauts at Esalen liked to claim, but it's there.

So, you wake up one morning and realize that you're not the person you wanted to be. You're someone else. You wonder how it happened. You look back upon your life, which appears to you as a caterpillar-track of choices you've made, most of which seemed right at the time – perhaps even inevitable – but that, one by one, have narrowed your corridor. And now you're twenty-eight or thirty-five or forty-seven or fifty-two or sixty-eight and you find yourself a very particular kind of person, with a very particular set of strengths and flaws, a particular worldview and a particular palette of moods and responses. When you were young, it seemed as if the future was as open as the prairie, that you could roam wherever you wanted, be whoever you decided. But what you've turned into is something like your mother, something like your father, but with a new layer on top that makes you feel more modern, more 'switched on', perhaps even a little bit smarter. If you were feeling especially dour, and if this morning happened to be notably gun-grey and mizzling, you might conclude that you're nothing more than a neoliberal update of your parents. This might be going too far. But it does suggest that your choices might not have been as free as they'd felt; that they'd been made under a heavy yet invisible genetic influence; that your soul is possessed by dead ancestors, who, working in conspiracy with the ghosts of your culture, mostly win.

It was a while before the shock of all this wore off. But when it did I was surprised to realize I actually found it deeply comforting; liberating, even. 'We have this culture in which you beat yourself up if you're not perfect,' said Daniel. 'We endorse the supposed effectiveness of therapy. Everyone thinks they can sort themselves out. But I think there's a naivety about that. If you're a more introverted person, or more neurotic, let yourself off the hook. Give yourself a break. It's bad enough, all this shit we have to take responsibility for, without also taking responsibility for being screwed up. I'm not advocating the kind of fatalism of, "I find it difficult getting to work on time so I'm not going to try." It's worth getting to work on time. But imagining you're going to be the kind of person who's going to get up three hours before work and bake wholemeal bread is kind of senseless.'

After reading Daniel's book, I began to see humans as belonging to a set of distinct sub-species, sort of genotribes. If someone was rude I'd think, "Oh dear, low agreeable," or if they were unpleasantly chirpy I'd think, "Save me from the extraverts." As Daniel finished his salad I became curious about his genotribe. 'Oh, I'm off-the-scale neurotic,' he told me.

'Would you describe yourself as an unhappy person?'

'Tremendously!' he smiled. People sometimes assume he has a wonderful life swanning around with his books and his teaching. 'But it's just a nightmare. I'm already worried about the next thing but three. In research, when a grant application fails, which it usually does, it's absolutely devastating. You just get so depressed. But other people go, "Oh well, I think they quite liked it. I'll stick it in next year." In a way, though, constant dissatisfaction with what you've done is really good because it makes you think, "Maybe there's a better way."' Being prone to worry and self-doubt can be an aid for success, he explained, 'but it's not pretty. We have this very foolish idea that people who are successful are happy and people who are happy are successful. I don't think either of those two things hold. It's a great hubris to imagine we have any idea what it's like to be anyone else. It's just foolishness when we say, "Oh, it must have been great to have been Mozart. All that music." It might've been a total nightmare to be Mozart. He might've been tormented. If someone says, "Oh, I wish I was Will Storr because he's written all these books," how do they know what it's like?'

'Yes, I suppose,' I said.

'But if you were the kind of person who was happy and satisfied, you probably wouldn't have written them,' he said. 'This is the paradox. This is why you've got to be very careful what you wish for. Would I like to be a less neurotic person? Well, no, because I probably wouldn't have done the most interesting things in my life. I mean, yes, I would've had an easier time of it. But you have to ask yourself, what do you value? Do you value being neurotic but doing things that, when you're on your deathbed, were actually quite interesting? Or do you value, I enjoyed it?'

I looked at him, expectantly, over our empty plates.

'There's no answer to that question,' he said.

*

Towards the end of my lunch with Daniel, he made a casual aside that changed the way I thought about myself, probably for ever. We were talking about the lingering effects of the self-esteem movement. 'We tend to lionize self-esteem,' he said, as we waited for the bill, the students hurrying out around us. 'We do this "people need to improve their self-esteem" thing endlessly, especially with kids. I've seen it in schools around here. But actually people with high self-esteem are pretty insufferable. It's all just very simplistic.'

'So having high self-esteem is not an unmitigated good?' I said.

'That's right. Absolutely.' He thought for a moment. 'But there are people who have low self-esteem who have a lot of problems and could benefit from it. Low self-esteem is almost synonymous with high neuroticism. The two go very, very strongly together.'

That was me, of course. I was a high neurotic and, as I'd been learning, neuroticism is a stable personality trait. It wasn't as if I was suffering from some curable malady, like my self had caught the flu, as the Vasco doctrine had it. It was who I was. I remembered what the sociologist John Hewitt had written about the self-esteem 'myth' being 'a contemporary tale in which men and women overcome mainly psychological obstacles to success and happiness'. Well, I'd fallen for that myth, utterly. I felt angry about it. I felt deceived.

What was remarkable was the extent to which I'd internalized these ideas. It was everywhere, during my teenage years, and my self had accepted Vasco's fantasy and used it, magpie-like, to build my narrative identity. For reasons I'd suspected were to do with my family environment, and Catholicism, I was a person with low self-esteem and one of my missions in life was to raise it. But, more damaging than that, I'd also absorbed my culture's conception of the perfect self. What a person *should* be was friendly, happy, popular, confident, comfortable with itself and with strangers. What a person *shouldn't* be was me. I'd compared my real self to the fantasy model and decided I was getting it wrong. I'd tried to change what couldn't be changed and berated myself for failing.

In doing this, I was hardly alone. Western culture prefers us not to believe we're defined or limited. It wants us to buy the fiction that the self is open, free, nothing but pure, bright possibility; that we're all born with the same suite of potential abilities, as neural 'blank slates', as if all human brains come off the production line at Foxconn. This seduces us into accepting the cultural lie that says we can do anything we set our minds to, that we can be whoever we want to be. This false idea is of immense value to our neoliberal economy. The game it compels us to play can best be justified morally if all the contestants start out with an equal shot at winning. Moreover, if we believe we're all the same, this legitimizes calls for deregulated corporations and smaller government: it means that the men and women who lose simply didn't want it badly enough, that they just didn't *believe* – in which case, why should anyone else catch their fall?

Exacerbating all this is the fact that we appear, like Freud, to be naturally wired to assume the minds of others are just like ours. This conspires to lead us further into the trap of judging others harshly. 'If I can do it, why can't they?' we think. And then, of course, we turn the logic in on ourselves, repeating that perfectionist mantra, 'If they can do it, why can't I?' All of this is compounded yet further by the fact we tend to radically overestimate our powers of control. Even if we decide to overrule the experts and assert we do have complete free will, it's been shown that Westerners are often more 'delusional about choice' than others, and likely to assume a person's failures are due to faults in the self, rather than biology, environment or situation.

Individualism makes us a blameful people. For us, blame is an object in space, a thing that exists, that belongs to someone. When we decide that it's ours, or somebody else's, we act in ignorance of the impossibly complex nature of why anybody behaves as they do. Of the addicts, the homeless, the violent, the obese, of those whose circumstances lead them into the utter darkness of prison, we're quick to condemn and slow to forgive. If a thing is what it is, then the thing you are, when you fail, is *bad*.

It's easy to see why we might be unconsciously drawn to this idea. It's been in the air since at least 500 BC, when the Greeks formed the individualist view of the perfectible self that still entrances us today. Bewitched by our culture, we expect others to be rational and always in control of their behaviour. We respond with disbelief and outrage when they disappoint. But people aren't what we think they are. We're not all constructed from the same precision-tooled machine parts. We haven't all been equally, perfectly designed to face the challenges of our environment. We're lumps of biology, mashed and pounded into shape by mostly chance events. Our 'human potential' is limited.

But this isn't the model of self that our culture keeps showing us. Instead, we're presented with an individual who has total free will and an ability to become whoever they choose. And who they usually choose to be is an extraverted, slim, individualistic, optimistic, hard-working, popular, socially aware yet high-self-esteeming individual with entrepreneurial guile – all characteristics Ayn Rand would've recognized as heroic. Because they're Greek, these cultural heroes will be beautiful both inside and out, and moving in the direction of perfection. Because they're Christian, they'll have clean and goodly interior selves. Because they're humanistic, they'll be authentic and 'real' and take responsibility for everything that happens to them. Because they're neoliberal, they'll be self-sufficient and successful and following their dreams with a ferocious hunger.

Neoliberal culture glorifies a kind of storyfied, dreamy-lensed version of this person. It creates, and then sells us, its bespoke hero. The stories we hear, whether they come from friends, from the newspapers, from fiction or from screen, often hail this category of person whilst diminishing others. The 'quiet loner' is practically synonymous with 'spree killer' in our culture, whilst in her book *Quiet* Susan Cain has famously described the 'extrovert ideal' it has a bias towards. Daniel Nettle writes that extraverts are 'highly active people who can lay their hands on vast stores of energy in pursuit of goals', are 'prepared to work very hard in pursuit of fame or money' and enjoy 'gaining status, receiving social attention'. He could be describing an ideal of self from Ancient Greece or a neoliberal hero.

As we've learned, it's not a coincidence that this model of ideal self also happens to be the one best equipped to get along and get ahead in the age of perfectionism – this era of heightened individualism, of financial crisis, of rising inequality, of personal debt, of small state, of deregulation, of austerity, of gig economy, of zero-hours contracts, of perfection-demanding gender ideals, of declining wages, of unrealistic body-image goals, of social media with its perfectionist presentation and its tribal outrage and demands for public punishment. These are the kinds of people who'll be more likely to win at the game that has been made of our world, the kinds who'll find a place in the boardroom or found billion-dollar hedge funds or start-ups – and then become powerful consumers, feeding back into the machine. This is who our modern tribal environment wants us to be.

One of the most surprising things I've come to realize on this long and startling journey is that storytelling is a form of tribal propaganda. Just as our hunter-gatherer ancestors' gossip about selfish and selfless people helped control the tribe, by teaching its members to be the selfless heroes, rather than the beaten, ostracized villains, so the same mechanisms exert powerful social pressures today. The stories of our neoliberal tribe insidiously persuade us that there's an ideal form of self and then defines it for us. We internalize this story and this hero. We make our tribe's story *our* story. We spread it around, in our own gossip and storytelling, becoming unconsciously complicit in the conspiracy. And then we try to become this hero – forcing ourselves into its shape, in the gym, in the office, on the therapist's couch. All too often, we fail. When the plots of our lives stall – when they 'fall severely short of standards and expectations' – and we can't see a way back to feeling heroic, then dangerous perfectionistic thinking can be triggered. We decide that we're losers at the game. We feel self-loathing. We might even find ourselves contributing to the already terrible statistics on suicide, self-harm and eating disorder. But what these stories don't tell us is that it's all a lie. None of us are heroes, not really. We're just us.

This isn't a message of hopelessness. On the contrary, what it actually leads us towards is a better way of finding happiness. The first step is to stop believing the tribal propaganda. Once you realize that it's all just an act of coercion, that it's your culture trying to turn you into someone you can't really be, you can begin to free yourself from its demands. Since I learned that low agreeableness and high neuroticism are relatively stable facets of my personality, rather than signs of some shameful psychological impurity, I've stopped berating myself so frequently. My head is now a much calmer place to be. I've even, perhaps ironically, become happier.

I'm not encouraging the popular, Fritz Perls-ish 'this is me, being real, deal with it' mode of self-acceptance. It's more a quiet understanding, an awareness that the way I am occasionally causes offence, and compensating, sometimes apologizing, but also trying not to attack myself for being myself. Of course, the culture we're raised in can't help but become us, to a great extent. We're never going to be able to fully deafen ourselves to its demands. There's a part of me that will always wish I had more friends, and was wealthier, thinner and more charming. But simply knowing 'perfect' is an illusion, that it's tribal propaganda, has been a deep comfort.

Stopping the war of perfection that's happening in your head is just the first step. Once you've quit trying to be who you're not, you can make an assessment of the things you're doing with your life. Professor Little, in his study of 'personal projects', argues that it's crucial to understand our limitations so that we can pursue goals that recognize them. 'We need to invest in those projects that bring meaning to us but which are achievable,' he told me. 'And if they're not achievable, then you don't just give up and say, "Life's wretched," you think of other ways of having a meaningful life. Change your goals.

What he's describing is a way of changing how we feel without having to change who we are. This is the way I've come to think of it: put a lizard on an iceberg and it's miserable. Now release it into the Sahara. The core of who that lizard is hasn't altered in the slightest and yet it's become happy. Its experience of life, but not its self, has transformed. If we want to inch towards happiness, then, we should stop trying to change ourselves and start trying to change our environment – the things we're doing with our lives, the people we're sharing it with, the goals we have. We should find projects to pursue which are not only meaningful to us, but over which we have efficacy. It doesn't have to be our job, it doesn't have to be something grandiosely altruistic, it just has to be *us*.

If the self is a story, then the story the Western self wants to tell is one of progress. Reality is chaos, chance and injustice, our future is illness, bereavement and death. All about us there is terrifying change, and there's little we can do to manage it. But our sense of self hides this disturbing fact from us. It leads us to believe we're heroes, captaining the plots of our lives. We're John Pridmore fighting Satan, we're Fritz Perls fighting Freud, we're Ayn Rand fighting the altruists, we're John Vasconcellos fighting the media cynics, we're Dan Faber fighting to one day dig the iron out of asteroids. It's when we lose the fights of our lives and keep on losing that we become stuck and humiliated, broken heroes, enemies of our evermore demanding culture. Then the story that is self starts to fail. It begins to creak and crack as the actual truth of what it is to be a living human presses in on it.

All we ever wanted was the illusion of control. But we have none, not really. And neither do the people around us who seem so intimidating in all their radiant perfection. Ultimately, we can all take comfort in the understanding that they're not actually perfect, and that none of us ever will be. We're not, as we've been promised, 'as gods'. On the contrary, we're animals but we think we're not animals. We're products of the mud.

*

Before I left Silicon Valley, I accompanied some residents of the Rainbow Mansion to a rocket launch. Elon Musk's company SpaceX had been contracted to take a NASA satellite into orbit. As we drove south out of Cupertino, I watched as the blue dot on my smartphone's map passed Big Sur and Esalen, not far to the west. With the highway running into the great Californian sky in front of us, I thought about the other journey I'd been on, which was now, finally, drawing to a close.

It had followed the passage of one idea – that of power being centred on the individual, in denial of other forces – as it had enchanted an entire people for 2,500 years. It had tracked this idea as it had shifted its shape in concert with great movements in the economy, and shown how these shifts had been codified, modified and evangelized by a chain of extraordinary people. And all the while, the women and men who'd been carried along inside this amorphous culture had been changed along with it.

I thought, once again, about how counterintuitive it all was; about how, no matter how convincing it might seem that our perspectives and beliefs come from a personal place of freely willed wisdom, my investigation hinted at the extraordinary extent to which we are, in fact, our culture. Which is not to say we're all clones, of course. We have different personalities, different in-group identities, different political biases, and so on. But all of that still sits within this dense web of stories, heroes, dreams and dreads that makes us all, no matter how far apart we might sometimes feel, a family.

When we arrived at the naval site, near Santa Barbara, where the event was taking place, Jeremy, Vanesa, Cate and I got out of the car and passed an image of the rocket on a video screen. It reminded me of the arrow tattoo I'd seen on CJ's forearm. It was a reference to *The Hunger Games*, of course, the story that had captured the imagination of her generation, in which a plucky member of the lower orders outsmarts the rules of the brutal competition she's been forced into and eventually conspires to topple the corrupt, dominating powers: 'The main character says, "When I lift my bow up and pull my arrow back there's only one way the arrow can go and that is forward,"' CJ had told me. 'I've always loved that. Go forward. Don't look back, no matter which direction life seems to be taking you.'

As we walked towards the viewing area, I overheard Vanesa talking to Jeremy. She was discussing her attempts at becoming an astronaut. One component of her NASA application had been a mental health check. I wanted to know about the things they'd asked. What sort of questions does an organization such as NASA believe define a person's sanity?

'One question was, do you think people are good or bad?' she said.

'And what did you say?'

'I think they're good.'

We walked on. I was grateful to be with them. I was having fun.

The fog was thick that morning. When the countdown began, everyone fell silent. We looked west, towards the cliffs that fell into the Pacific Ocean, our mobile phones in hand. A great roaring came from that place. Even though we couldn't see what was coming, we smiled. We raised our cameras. The ground shook.

Acknowledgements

Thanks to my editor, Kris Doyle, who I exhausted on more than one occasion during the making of this book (I hereby promise to never experiment with a no-chapters, stream of consciousness narrative ever again), and to Paul Baggaley, Charlie Campbell, Will Francis, Nicholas Blake and Greg Callus.

Thanks also to Bobbie Johnson of *Matter*, who commissioned the profile of Roy Baumeister that inspired this book (and to Roy Baumeister for putting up with endless hours of interviews), to the excellent Dr Tasha Eurich for your invaluable comments, to Amelia Jean Jones for your lifesaving transcription services, to Craig Pearce for your friendship and counsel, to Mun Keat Looi, Alex Bilmes, Matthew Drummond, Kirby Kim, Rebecca Folland and also to my panel of brilliant brains for reading and checking the manuscript, in whole or in part, prior to publication. They are: Professor Sophie Scott, Professor Constantine Sedikides, Dr Stuart Ritchie and Professor Helen Morales. Thanks, too, to the extremely kind (empathetic?) Professor Paul Bloom, who went above and beyond by helping me, a complete stranger, ferret out a pesky missing citation on Christmas Eve.

Most thanks of all, though, go to the many people who have given their time and thoughts to me, either in person or on the telephone. To all those who appear in these pages, I can't thank you enough.

Finally, thanks to my amazing and brilliant wife, Farrah.

A note on my method

As a journalist, my knowledge is broad but shallow. Most of the ideas covered in this book are well documented in a wide range of excellent books and periodicals – both academic and popular – which were my principal source for research. I supplemented this research with interviews with experts, generally on the professor level. Most of the general concepts in this book are relatively uncontroversial and broadly accepted.

In the few areas where I explored more controversial science, I sought expert counsel where those studies threatened to be too complex for a lay journalist to appropriately understand.

Finally, I recruited a team of academics with appropriate specialisms to read through the manuscript either in whole or in part. They offered notes and advice where I had erred.

This is an imperfect system as it relies on many secondary sources. It also centres on many findings from social psychology, a field which is undergoing an unsettling amount of churn at the moment. Moreover, I do not declare myself to be free of the biases that afflict any writer, and I'm certainly not immune to making mistakes. If any errors are noted, or if new findings supersede claims made in the text, I would be very grateful to receive notification via my website, willstorr.com, so that future editions of *Selfie* can be corrected and updated.

Naturally, this book contains only a fraction of a fraction of the relevant science. Other academics will, surely, disagree with those whom I quote in these pages. If any of it piques your interest, I urge you to dig deeper, where you will no doubt find science that is newer and in conflict with some of the work here.

Some names have been changed, all interviews are edited, ellipses are not used within hybrid quotes, which are used in the interests of concision. Several sections of this book have appeared previously, in different forms, in periodicals including *Mosaic*, *Esquire*, the *New Yorker*, *Matter*, the *Guardian*, *Aeon*, *GQ Australia* and *Cosmopolitan*.

Notes and references

Book Zero: The Dying Self

page

3 **A few days previously, Debbie:** My account of Debbie's attempted suicide is based on an interview with Debbie, and the account she has given in her self-published memoir, *Sex, Suicide and Serotonin.*

6 **overall rates in the US and UK have seen a general decline since the 1980s:** Office for National Statistics: Suicides in the United Kingdom: 2014 registrations.

6 **the introduction of 'blockbuster' antidepressants** Gauging an accurate view on this is hard for a variety of reasons: firstly, for a minority of patients, a side effect of taking them is thought by some to be increased suicidality (although some studies don't find this effect); secondly, researchers are currently at odds on their overall impact on suicides – some deny any effect, whilst others find they have caused a highly significant decrease.

7 **more people die by suicide than in all the wars, terrorist attacks, murders and government executions combined:** WHO Global Health Observatory Repository, apps.who.int/gho/data.node.main.RCODWORLD?lang=en, accessed 7 September 2015.

7 **. . . *twenty times* more attempted:** American Foundation for Suicide Prevention: https://afsp.org/about-suicide/suicide-statistics/. The AFSP estimate the true number to be higher. Gordon Flett had it at twenty-five attempts to every completion. The WHO report 'Preventing Suicide, A Global Imperative' estimates the figure at twenty (p. 26) but acknowledges the data is far from perfect.

8 **As it is, men make up around 80 per cent**: In the UK it's 78 per cent: 'Male suicide rate worst since 2001, ONS reveals: Office for National Statistics reveals male suicide rate in UK has "increased significantly" since 2007, while female rates have stayed "consistently lower"', *Daily Telegraph,* 19 February 2015. In the US, it's 79 per cent: CDC, National Center of Injury Prevention and Control, Suicide fact sheet, 2012. In Australia it's 77 per cent: Australian Bureau of Statistics, Gender Indicators, Australia, January 2013, Suicides. In Canada it's 'about three times that of

women.': 'The silent epidemic of male suicide', Dan Bilsker and Jennifer White, *British Columbia Medical Journal* (December 2011), vol. 53, no. 10, pp. 529–34.

8 **There are many vulnerabilities:** Rory O'Connor.

9 **One examination of:** 'Gender-Related Schemas and Suicidality: Validation of the Male and Female Traditional Gender Scripts Questionnaires', Martin Seager, Luke Sullivan, John Barry, *New Male Studies: An International Journal* (2014), vol. 3, issue 3, pp. 34–54.

10 **. . . a report on male suicide that Rory co-authored:** 'Men, Suicide and Society, Why disadvantaged men in mid-life die by suicide', Samaritans' research report, Clare Wyllie et al., September 2012.

10 **Baumeister theorized:** 'Suicide as escape from self', Roy Baumeister, *Psychological Review* (January 1990), 97(1), pp. 90–113.

11 **Although the worldwide data is relatively scant:** Rory O'Connor, comment during fact check, via email.

11 **. . . women actually attempt suicide in greater numbers:** 'Cross-national prevalence and risk factors for suicidal ideation, plans and attempts', Matthew K. Nock et al., *British Journal of Psychiatry* (January 2008), 192(2), pp. 98–105.

12 **some of which display 'triple zero' mannequins:** 'Tiny waist, insect legs: fashion still in thrall to triple zero', Josh Boswell and Elisabeth Perlman, *Sunday Times*, 25 October 2015.

12 **. . . 61 per cent of young women and girls in the UK felt happy:** 2016 Girls' Attitudes Survey: https://www.girlguiding.org.uk/globalassets/docs-and-resources/research-and-campaigns/girls-attitudes-survey-2016.pdf.

12 **self harm . . . eating disorders . . . have perfectionism as a predictor:** 'Predicting depression, anxiety and self-harm in adolescents: The role of perfectionism and acute life stress', Rory O'Connor et al., *Behaviour Research and Therapy* (January 2010), 48(1), pp. 52–9. Here, the researchers found evidence that social perfectionism 'interacted with acute life stress to predict self-harm.' 'Perfectionism and eating disorders: Current status and future directions', Anna M. Bardone-Cone et al., *Clinical Psychology Review* (April 2007), 27(3), pp. 384–405.

12 **risen by around 30 per cent:** 'Perfectionism Is Increasing Over Time: A Meta-Analysis of Birth Cohort Differences From 1989 to 2016', T. Curran and A. P. Hill, *Psychological Bulletin* (2017). Advance online publication, full study supplied by the authors.

12 **The number of adults reporting self-harm between 2000 and 2014**

has more than doubled: *Adult Psychiatric Morbidity Survey: Survey of Mental Health and Wellbeing, England, 2014*, NHS Digital, Chapter 12: 'Suicidal thoughts, suicide attempts and self harm'.

12 **One senior psychiatrist told reporters that rises in youth self-cutting:** 'NHS figures show "shocking" rise in self-harm among young', Denis Campbell, *Guardian*, 23 October 2016.

12 **in the US anxiety and depression has been rising in adolescents since 2012:** 'Teen Depression and Anxiety: Why the Kids Are Not Alright', Susanna Schrobsdorff, *Time Magazine*, 27 October 2016.

12–13 **Dr Jackie Cornish, of NHS England, said, 'In common with most experts':** 'NHS figures show "shocking" rise in self-harm among young', Denis Campbell, *Guardian*, 23 October 2016.

13 **Paediatrician Dr Colin Michie placed much:** 'Stark rise in eating disorders blamed on overexposure to celebrities' bodies', Denis Campbell, *Guardian*, 25 June 2015.

13 **One US study found body dysmorphic disorder:** 'The prevalence of body dysmorphic disorder in the United States adult population', L. M. Koran et al., *CNS Spectrums* (April 2008), **13**(4), pp. 316–22. See also a 2016 study that found body dissatisfaction to be nearly as prevalent. 'Correlates of appearance and weight satisfaction in a U.S. National Sample: Personality, attachment style, television viewing, self-esteem, and life satisfaction', David Frederick et al., *Body Image* (June 2016), vol. 17, pp. 191–203. (15 per cent of men and 20 per cent of women were very to extremely dissatisfied with their weight.)

13 **'muscle dysmorphia', a condition . . . it's thought that up to four million Americans:** 'Body Image Disorders and Abuse of Anabolic-Androgenic Steroids Among Men', Harrison G. Pope Jr et al., *JAMA* (January 2017), **317**(1), pp. 23–4.

13 **43 per cent jump** Laure Donnelly, 'Number of men referred for eating disorder treatment rises by 43 per cent', *Daily Telegraph*, 24 July 2017.

13 **needle exchanges in some cities:** 'Spiralling anabolic steroid use leaves UK facing health time bomb, experts warn', Peter Walker, *Guardian*, 19 June 2015.

13 **Spending on gym membership in 2015 alone rose by 44 per cent:** 'UK gym membership spending up by 44%', Rebecca Smithers, *Guardian*, 18 August 2015.

13 **A government enquiry into the problem:** All Party Parliamentary Group on Body Image, final report, 19 September 2014, p. 7.

13 **When a University of Pennsylvania task force:** 'Campus Suicide and the Pressure of Perfection', Julie Scelfo, *New York Times*, 28 July 2015.

14 **Gordon co-authored a paper:** 'The destructiveness of perfectionism revisited: Implications for the assessment of suicide risk and the prevention of suicide', Gordon L. Flett, Paul L. Hewitt, Marnin J. Heisel, *Review of General Psychology* (September 2014), **18**(3), pp. 156–72.

14 **The most comprehensive study** Christian Jarrett, 'Perfectionism as a risk factor for suicide – the most comprehensive test to date', *British Psychological Society Digest*, 27 July 2017.

15 **in July 2016, when sixteen-year-old Phoebe Connop:** 'Teenage girl killed herself amid fears she would be branded racist over joke photo she sent friends, inquest hears', Lydia Willgress, *Daily Telegraph*, 28 August 2016.

16 **A *New York Times* report into rising suicide rates:** 'Campus Suicide and the Pressure of Perfection', Julie Scelfo, *New York Times*, 28 July 2015.

16 **Take Meredith:** This name has been changed at the request of the interviewee.

17 **a major new study:** 'Perfectionism Is Increasing Over Time: A Meta-Analysis of Birth Cohort Differences From 1989 to 2016', T. Curran and A. P. Hill, *Psychological Bulletin* (2017). Advance online publication, full study supplied by the authors. NOTE: This study was published on 2 January 2018, seven months after the first publication of *Selfie*.

Book One: The Tribal Self

page

23 **John Pridmore sat square in front of me:** My recounting of John's story is taken from my interview and from his autobiography, *From Gangland to Promised Land*, with Greg Watts (2nd edn: Transform Management, 2004).

28 **Humans have spent more than 90 per cent of their time:** *Human Frontiers, Environments and Disease: Past Patterns, Uncertain Futures*, Tony McMichael (Cambridge University Press, 2004), p. 133.

28 **We share a common ancestor, and more than 98 per cent of our DNA:** *Why We Believe in God(s)*, J. Anderson Thomson Jr. with Clare Aukofer (Pitchstone Publishers, 2011), p. 35.

28 **Along with the bonobo:** *Demonic Males*, Richard Wrangham and Dale Preston (Bloomsbury, 1996), p. 26.

28 **if about two-thirds smaller than:** *Who's In Charge*, Michael S. Gazzaniga (Robinson, 2011), p. 149.
28 **They hide their feelings to get their own way:** *Demonic Males*, Richard Wrangham and Dale Preston (Bloomsbury, 1996), p. 24.
28 **They hold long-term grudges:** *Demonic Males*, Richard Wrangham and Dale Preston (Bloomsbury, 1996), p. 23.
28 **They negotiate peace by bringing aggressors together:** *Demonic Males*, Richard Wrangham and Dale Preston (Bloomsbury, 1996), p. 24.
28 **They have a sense of fair play:** *Just Babies*, Paul Bloom (Bodley Head, 2013), p. 80.
28 **and are motivated to punish the selfish:** *Moral Tribes*, Joshua Greene (Atlantic Books, 2013), p. 41.
28 **Weaker and younger chimpanzees regularly engage:** *The Origins of Virtue*, Matt Ridley (Penguin, 1996), pp. 157–9.
29 **They engage in political beatings and murders:** *Demonic Males*, Richard Wrangham and Dale Preston (Bloomsbury, 1996), p. 131.
29 **These acts of violence are not the product:** *Demonic Males*, Richard Wrangham and Dale Preston (Bloomsbury, 1996), p. 187.
29 **Alpha chimps studied by the famous primatologist:** *Our Inner Ape*, Frans de Waal (Granta, 2005), p. 43.
29 **De Waal's account of the behaviour:** *Our Inner Ape*, Frans de Waal (Granta, 2005), p. 44.
30 **An alpha male's reign usually lasts less than five years:** *Our Inner Ape*, Frans de Waal (Granta, 2005), p. 68.
30 **The biological anthropologist Professor Richard Wrangham has observed that:** *Demonic Males*, Richard Wrangham and Dale Preston (Bloomsbury, 1996), p. 26.
30 **humans, upon meeting someone new, will automatically:** 'Intergroup Relations', D. Messick, D. Mackie, *Annual Review of Psychology* (1989), vol. 40, pp. 45–81.
30 **babies universally prefer faces of their own race:** *Just Babies*, Paul Bloom (Bodley Head, 2013), p. 105.
30 **Children as young as six:** *Just Babies*, Paul Bloom (Bodley Head, 2013), p. 113.
30 **The effect of our tribal brains has been shown in numerous experiments:** For more information, see my previous book, *The Heretics* (Picador, 2013). One classic study of this effect is 'Experiments in intergroup discrimination', H. Tajfel, *Scientific American* (1970), vol. 223, pp. 96–102.

31 **we are great apes that sit in the primate superfamily Hominoidea:** http://australianmuseum.net.au/humans-are-apes-great-apes.

33 **Caring about what others think of us:** *The Self Illusion*, Bruce Hood (Constable, 2011), p. 138.

33 **Children start attempting to manage their reputations at around the age of five:** 'Five-Year Olds, but Not Chimpanzees, Attempt to Manage Their Reputations', J. M. Engelmann, E. Herrmann, M. Tomasello, *PLoS ONE* (October 2012), 7(10).

33 **Those who earned bad ones could easily be beaten, killed or ostracized:** Interview, Dr Giorgia Silani, International School for Advanced Studies.

34 **Studies that measure how much of human conversation it constitutes put the figure at between 65 and 90 per cent:** The 65 per cent figure comes from Professor Joshua Greene and is quoted in his book *Moral Tribes* (Atlantic Books, 2013), p. 45. The 90 per cent figure comes from Professor Michael Gazzaniga and is quoted in his book *Human* (Harper Perennial, 2008), p. 96.

34 **At the age of three, children start communicating:** 'Even preschoolers like to gossip', Christian Jarrett, *Psychological Research Digest*, 2 September 2016.

34 **Despite the gendered stereotype:** *Human*, Michael Gazzaniga (Harper Perennial, 2008), p. 96.

34 **A study of gossip at a Belfast school:** *Just Babies*, Paul Bloom (Bodley Head, 2013), p. 95.

34 **negative gossip can even affect our vision:** 'Bad gossip affects our vision as well as our judgment', Ed Yong, *Discover Magazine*, 20 May 2011.

34 **'Dunbar's number', as it's known:** http://webpages.charter.net/sn9/literature/neocortex.html.

34 **would've generated powerful feelings of moral outrage:** Jonathan Haidt has described gossip as both 'teacher and policeman', *Human*, Michael Gazzaniga (Harper Perennial, 2008), p. 96. There is debate as to whether altruistic punishment is an evolved instinct (see *Just Babies*, Paul Bloom (Bodley Head, 2013), ch. 3), although Bloom writes, 'Humans everywhere do punish free riders', p. 91.

35 **in the well-known words of the psychologist Professor Robert Hogan:** *The Redemptive Self*, Dan P. McAdams (Oxford University Press, 2013), p. 29.

35 **'are not acquired through learning':** *Just Babies*, Paul Bloom (Bodley Head, 2013), p. 8.

36 **a simple puppet show:** *Just Babies*, Paul Bloom (Bodley Head, 2013), p. 27.

36 **A large body of such work now suggests:** For a brilliant account of this work, I recommend Joshua Greene's *Moral Tribes* (Atlantic Books, 2013). He describes the essence of what we consider a moral person to be as 'altruism, unselfishness, a willingness to pay a personal cost to benefit others', p. 23.

36 **The underside of that, of course, is selfishness:** There's currently disagreement about whether tribal punishment of the selfish came about because it benefited the group, the individual, or both. Whatever the outcome of the group-selection vs individual-selection debate, the critical point in this context is simply that selfishness was seen as bad and such behaviour was penalized. Whether selfish acts were bad for me or bad for everyone, or both, the tribe could only function smoothly if routine attempts were made to suppress them.

36 **toddlers naturally expect sharing between members of their group:** 'Naïve theories of social groups', M. Rhodes, *Child Development* (2012), 83(6), pp. 1900–16.

36 **They even keep track of the politics of sharing:** 'It's payback time: Preschoolers selectively request resources from someone they had benefitted', Markus Paulus, *Developmental Psychology* (August 2016), 52(8), pp. 1299–1306.

36 **When offered a smaller portion of sugary treats:** *Just Babies*, Paul Bloom (Bodley Head, 2013), pp. 80–81.

36 **What we're doing is trying to control the selfishness:** According to Jonathan Haidt: 'Moral systems are interlocking sets of values, virtues, norms, practices, identities, institutions, technologies, and evolved psychological mechanisms that work together to suppress or regulate selfishness and make social life possible.' *Who's In Charge*, Michael S. Gazzaniga (Robinson, 2011), p. 166.

37 **They *weren't* surprised:** 'Naïve theories of social groups', M. Rhodes, *Child Development* (2012), 83(6), pp. 1900–16.

37 **Selfless acts are most often made *on behalf of our people*:** Altruistic acts, at least according to some social scientists, mostly tend to happen when the costs to the individual aren't too grave. Acts of kindness towards those outside the tribe or ingroup are relatively rare. And of course, when we do behave altruistically, we also get rewarded with an enhanced reputation. For more, see *Moral Tribes*, Joshua Greene (Atlantic Books, 2013), p. 39: 'We care most of all about our relatives and friends, but we also care about

acquaintances and strangers . . . We're willing to help strangers, expecting nothing in return, so long as it's not too costly.'

37 **The celebrated mythologist Joseph Campbell:** *The Power of Myth*, Joseph Campbell with Bill Moyers (Broadway Books, 1998), p. 127.

38 **We begin to experience guilt before the age of one:** *Just Babies*, Paul Bloom (Bodley Head, 2013), p. 55.

44 **first properly exposed in the 1960s by a team that included cognitive neuroscientist:** My account of Gazzaniga's confabulation experiments is sourced from his books *Who's In Charge?* (Robinson, 2011) and *Human* (Harper Perennial, 2008). Another excellent telling can be found in *The Happiness Hypothesis*, Jonathan Haidt (Heinemann, 2006).

46 **My favourite involves people who were shown:** *Mindwise*, Nicholas Epley (Allen Lane, 2014), p. 31.

46 **Our brains invent stories like this because:** *The Self Illusion*, Bruce Hood (Constable, 2011), p. 114.

46 **'When we set out to explain our actions . . . listening to people's explanations for their actions is interesting . . .':** *Who's In Charge?*, Michael Gazzaniga (Robinson, 2011), p. 77.

47 **'If you devote your time to . . .':** 'The Danger Of Inadvertently Praising Zygomatic Arches', Robert Sapolsky, *Edge*, 2013.

47 **Most of those who argue that we *do* have free will:** For various fascinating discussions on the issue, see Bruce Hood's *The Self Illusion* (Constable, 2011), David Eagleman's *Incognito: The Secret Lives of the Brain* (Canongate, 2011) and Jonathan Haidt's *The Happiness Hypothesis* (Heinemann, 2006).

Book Two: The Perfectible Self

page

54 **The scholar Professor Werner Jaeger has written of *kalokagathia*'s roots:** *Paideia, The Ideals of Greek Culture*, vol. 1, Werner Jaeger, trans. Gilbert Highet (Oxford University Press, NY, 1945), p. 4.

55 **the weight of a child's brain increases by more than 30 per cent:** *The Self Illusion*, Bruce Hood (Constable, 2011), pp. 14–15.

55 **By the age of two, a human will have generated over a hundred trillion synapses, double:** *The Brain*, David Eagleman (Pantheon Books, 2015), Kindle location 85.

55 **Six-month-olds can recognize the faces of individuals from other races:** *The Self Illusion*, Bruce Hood (Constable, 2011), p. 28.

55 **They can even readily identify *monkey* faces:** *The Self Illusion*, Bruce Hood (Constable, 2011), p. 27.

55 **Babies can hear tones in foreign languages:** *The Brain*, David Eagleman (Pantheon Books, 2015), Kindle location 85.

55 **They're also thought to experience synaesthesia:** 'Synesthesia: A new approach to understanding the development of perception', Ferrinne Spector, Daphne Maurer, *Developmental Psychology* (January 2009), 45(1), pp. 175–89.

55 **These connections start dying off at a rate of up to 100,000 per second:** *The Self Illusion*, Bruce Hood (Constable, 2011), p. 15.

56 **In a major study, researchers in Queensland collated the results of 2,748 papers:** 'Meta-analysis of the heritability of human traits based on fifty years of twin studies', Tinca J. C. Polderman et al., *Nature Genetics* (May 2015), 47, pp. 702–9.

56 **Co-author Beben Benyamin added:** 'Are we products of nature or nurture? Science answers age-old question', 19 May 2015, https://www.theguardian.com/science/2015/may/19/are-we-products-of-nature-or-nuture-science-answers-age-old-question.

58 **Children growing up in Tanzania:** Interview, Sophie Scott.

59 **to get along, which gives us prestige, and get ahead, which gives us status:** Fact check notes, Constantine Sedikides.

59 **around 45,000 years ago:** *The Domesticated Brain*, Bruce Hood (Penguin, 2014), p. 45.

59 **just a fifth of Ancient Greece was well suited for agriculture:** *Ancient Greece* (3rd ed), Sarah B. Pomeroy et al. (Oxford University Press, 2012), pp. 18–19.

60 **'We dwell about the sea like ants or frogs around a pond':** *Phaedo*, Plato, 109a-b.

60 **They were pirates:** *The Geography of Thought*, Richard E. Nisbett (Nicholas Braeley, 2003), p. 34.

60 **Their ports:** 'Classics for the people – why we should all learn from the ancient Greeks', Edith Hall, *Guardian*, 20 June 2015.

60 **'appears to be the beginning of a new conception of the value . . .':** *Paideia, The Ideals of Greek Culture*, vol. 1, Werner Jaeger, trans. Gilbert Highet (Oxford University Press, NY, 1945), p. xix.

60 **a 'civilization of cities':** *Ancient Greece*, Paul Cartledge (Oxford University Press, 2011), p. 2.

60 **composed, chiefly, of more than a thousand:** *Ancient Greece*, Paul Cartledge (Oxford University Press, 2011), p. 1.

60 **In faraway lands kings and tyrants claimed:** *Ancient Greece* (3rd ed), Sarah B. Pomeroy et al. (Oxford University Press, 2012), p. 23.

60 **'You started your speech with a false statement . . .':** *The Suppliants*, Euripides, 103–8.

61 **such as that of the 'father of comedy':** *Aristophanes in Performance 421 BC–AD 2007: Peace, Birds and Frogs*, Edith Hall and Amanda Wrigley (Oxford, 2007), p. 1.

61 **Athenians were free to travel great distances:** *The Geography of Thought*, Richard E. Nisbett (Nicholas Braeley, 2003), p. 2.

61 **A commoner could debate royalty:** *The Geography of Thought*, Richard E. Nisbett (Nicholas Braeley, 2003), p. 3.

61 **If a person clashed with their neighbours:** *The Geography of Thought*, Richard E. Nisbett (Nicholas Braeley, 2003), p. 30.

61 **The ordinary Greek would seek to control the gods:** *Ancient Greece* (3rd ed), Sarah B. Pomeroy et al. (Oxford University Press, 2012), p. 84.

61 **The closest thing they had to heaven . . . hell:** *Classical Mythology*, Helen Morales (Oxford University Press, 2012), p. 43.

61 **'The Greeks, more than any other ancient peoples . . .':** *The Geography of Thought*, Richard E. Nisbett (Nicholas Braeley, 2003), pp. 2–3.

62 **Historian Adrienne Mayor writes that:** 'Megafauna', *London Review of Books*, vol. 37, no. 13, p. 25.

62 **It's for this reason, argues Jaeger:** *The Ideals of Greek Culture*, vol. 1, Werner Jaeger, trans. Gilbert Highet (Oxford University Press, NY, 1945), p. xxii.

63 **Skills in debating, which could take place anywhere:** *Ancient Greece* (3rd ed), Sarah B. Pomeroy et al. (Oxford University Press, 2012), pp. 2–3.

63 **'Potter resents potter . . .':** *Ancient Greece* (3rd ed), Sarah B. Pomeroy et al. (Oxford University Press, 2012), p. 79.

63 **For the victor not to be honoured:** *Ancient Greece* (3rd ed), Sarah B. Pomeroy et al. (Oxford University Press, 2012), p. 79.

63 **'the greatest of human tragedies':** *The Ideals of Greek Culture*, vol. 1, Werner Jaeger, trans. Gilbert Highet (Oxford University Press, NY, 1945), p. 4.

63 **For academics such as Nisbett:** *The Geography of Thought*, Richard E. Nisbett (Nicholas Braeley, 2003).

65 **Our 'episodic memory' . . . 'autobiographical memory':** Some consider episodic and autobiographical memory to be the same thing.

65 **To have a self is to feel as if we are:** *Making up the Mind*, Chris Frith (Blackwell, 2007), p. 109.

65 **The healthy, happy brain:** For more information on this section please see my previous book, *The Heretics.*

66 **'virtually all individuals irrationally inflated':** 'The Illusion of Moral Superiority', Ben M. Tappin and Ryan T. McKay, *Social Psychological and Personality Science* (2016), pp. 1–9.

66 **Work by psychologists including Professor Nicholas Epley:** *Mindwise*, Nicholas Epley (Penguin, 2014), p. 54.

66 **the stories our parents tell us, and their characteristic shape:** *The Construction of the Self*, Susan Harter (Guildford Press, 2012), p. 39.

67 **Between five and seven, the content:** *The Construction of the Self*, Susan Harter (Guildford Press, 2012), p. 50.

67 **during adolescence, according to the psychologist Professor Dan McAdams:** *The Redemptive Self*, Dan P. McAdams (Oxford University Press, 2013), p. xii.

67 **'Culture provides each person with an extensive menu of stories about how to live . . .' etc.:** *The Redemptive Self*, Dan P. McAdams (Oxford University Press, 2013), p. 284.

67 **Turning our lives into myth:** *The Stories We Live By*, Dan P. McAdams (Guilford Press, 1997), p. 91.

67 **Our story gives our life meaning and purpose:** For more information on this please see my previous book, *The Heretics.*

67 **'giving yourself to some higher end . . .':** *The Power of Myth*, Joseph Campbell with Bill Moyers (Broadway Books, 1998), p. 126.

68 **This incomplete creature is immensely powerful:** *The Seven Basic Plots*, Christopher Booker (Continuum, 2005), p. 555.

68 **'The point is that the disorder in the upper world cannot':** *The Seven Basic Plots*, Christopher Booker (Continuum, 2005), pp. 123–4.

68 **'Stories present us with an ideal picture of human nature':** *The Seven Basic Plots*, Christopher Booker (Continuum, 2005), p. 268.

69 **'a strong protagonist':** *Redirect*, Timothy D. Wilson (Penguin, 2013), p. 268.

69 **on great plains and amongst gentle mountains:** *The Geography of Thought*, Richard E. Nisbett (Nicholas Braeley, 2003), p. 34.

69–70 **rarely encountered foreigners or foreign beliefs:** *The Geography of Thought*, Richard E. Nisbett (Nicholas Braeley, 2003), p. 31.

70 **Master Kong, or Confucius, arrived:** My account of Confucius was mostly sourced from *Confucius and the World He Created*, Michael Schuman (Basic, 2015) and *Confucianism*, Daniel K. Gardner (Oxford University Press, 2014).

You don't have to go to China to see how individualist and collectivist selves combine to create different places for themselves to live in. Some of the effects were shown in a remarkable 1953 MIT study of two outwardly similar New Mexican towns of around 250 people that the sociologists named Homestead and Rimrock. Both towns were recognizably American in culture, but had slight shifts in emphasis when it came to how individualistic they were. Rimrock had been founded by Mormon missionaries and tended towards community co-operation whilst Homestead was settled by dustbowl migrants. With its holy potpourri of Christian denominations including Baptists, Presbyterians, Methodists, Nazarenes, Campbellites, Holinesses, Seventh Day Adventists, Mormons, Catholics and Present Day Disciples – there were even a couple of atheists – Homestead was much more individualistic. Their differences in worldview were relatively marginal, except when it came to their social relationships. From these differences sprouted significant real effects that deeply affected each town.

When the opportunity came to gravel the streets, a meeting was called in collectivist Rimrock in which, after disagreements, it was decided each family would contribute $20, with a trader paying more. The job was done. They had gravelled streets. When the same opportunity arrived in individualist Homestead, residents baulked at the idea of paying for gravel they might not directly benefit from. A few local businesses did pay for it to be laid just in front of their own properties, but that 'left the rest of the village streets a sea of mud in rainy weather'.

Something similar happened when plans for a high-school gymnasium were considered. In collectivist Rimrock, after another round of disagreement, it was decided each able-bodied man would contribute either fifty hours of labour or $50 to pay someone to do their share. The project was completed. They had a gymnasium. And yet, forty miles away in individualist Homestead, the same plan was rejected. When funding towards construction was sourced, everyone demanded payment by the hour. But then the cash ran out and work ceased. 'Today,' wrote the authors, 'a partially completed gymnasium, and stacks of some 10,000 adobe bricks disintegrating slowly with the rains, stand as monuments to the individualism of the Homesteaders.'

Of course, the extremely high suicide rates in East Asia and the bloody and miserable history of communism are enough to tell us that adopting a hard form collectivism won't magic you into some kind of utopia. But, limited to just one example as this study is, it's interesting in at least indicating how subtle shifts one way or the other can have significant effects.

71 **'Confucius expected people to do the right thing . . .':** *Confucius and the World He Created*, Michael Schuman (Basic Books, 2015), p. xvii.

71 **'Whenever a visitor wearing a Confucian hat comes to see him . . .':** *Confucianism*, Daniel K. Gardner (Oxford University Press, 2014), p. 5.

71 **Scholars such as Richard Nisbett argue:** *The Geography of Thought*, Richard E. Nisbett (Nicholas Braeley, 2003), pp. 31–2.

72 **Ninety-five per cent of modern China:** *The Geography of Thought*, Richard E. Nisbett (Nicholas Braeley, 2003), p. 31.

72 **This means that East Asians tend to be more aware of what's happening:** Interview, Richard E. Nisbett.

72 **Tests involving videos of fish:** 'Culture and cause: American and Chinese attributions for social and physical events', Michael W. Morris, Kaiping Peng, *Journal of Personality and Social Psychology* (December 1994), 67(6), pp. 949–71.

72 **the East Asians made over 60 per cent more references to objects in the background:** 'Attending holistically versus analytically: comparing the context sensitivity of Japanese and Americans', T. Matsuda and R. E. Nisbett, *Journal of Personality and Social Psychology* (November 2001), 81(5), pp. 922–34.

72 **Examinations of youngsters' drawings suggest:** 'Holistic Versus Analytic Expressions in Artworks: Cross-Cultural Differences and Similarities in Drawings and Collages by Canadian and Japanese School-Age Children', S. Senzaki et al, *Journal of Cross-Cultural Psychology* (June 2014), 45(8), pp. 1297–1316.

73 **Researchers deconstructed stories:** 'Culture and cause: American and Chinese attributions for social and physical events', Michael W. Morris, Kaiping Peng, *Journal of Personality and Social Psychology* (December 1994), 67(6), pp. 949–71.

74 **three communities in Turkey's Black Sea region:** 'Ecocultural basis of cognition: Farmers and fishermen are more holistic than herders', Ayse K. Uskul, Shinobu Kitayama, and Richard E. Nisbett, *Proceedings of the National Academy of Sciences of the United States of America* (June 2008), 105(25), pp. 8552–6.

74 **those from the southern states reacted more aggressively:** 'Insult, aggression, and the southern culture of honor: an "experimental ethnography"', D. Cohen, R. E. Nisbett, B. F. Bowdle, N. Schwarz,

Journal of Personality and Social Psychology (May 1996), **70**(5), pp. 945–60.

75 **A study led by Thomas Talhelm of the University of Virginia honed in on:** 'Large-Scale Psychological Differences Within China Explained by Rice Versus Wheat Agriculture', T. Talhelm et al., *Science* (May 2014), **344**(6184), pp. 603–8.

75 **Asians don't feel as in control of their lives:** *The Geography of Thought*, Richard E. Nisbett (Nicholas Braeley, 2003), p. 97.

75 **Amongst Chinese students . . . in industry . . . .:** *Quiet*, Susan Cain (Penguin, 2013), p. 187.

76 **In Chinese, there isn't a word for individualism:** Interview, Richard E. Nisbett.

76 **The term for 'human being' in Japanese and Korean:** Interview, Uichol Kim.

76 **Most studies show that East Asians have lower self-esteem:** 'Self-esteem in American and Chinese (Taiwanese) children', Lian-Hwang Chiu, *Current Psychology* (December 1992), **11**(4), pp. 309–13.

78 **according to Professor Qi Wang:** *The Autobiographical Self in Time and Culture*, Qi Wang (Oxford University Press, 2013), pp. 46 and 52.

79 **'Life is change that yearns for stability':** *The Cultural Animal*, Roy Baumeister (Oxford University Press, 2005), p. 102.

81 **South Korea has, by some counts, the second highest rate in the world:** 'Suicide in the World', Peeter Värnik, *International Journal of Environmental Research and Public Health* (March 2012), **9**(3), pp. 760–71.

81 **Around forty South Koreans take their own lives every day . . . five times higher:** 'Tackling South Korea's high suicide rates', Lucy Williamson, *BBC News*, 8 November 2011.

81 **One poll:** 'Poll Shows Half of Korean Teenagers Have Suicidal Thoughts', Yewon Kang, *Korea Realtime*, 20 March 2014.

82 **rates fell by an astonishing 58 per cent:** 'Back from the edge', *The Economist*, 28 June 2014.

83 **Sheldon Wolin writes:** *Politics and Vision: Continuity and Innovation in Western Political Thought* (Princeton, 2004), pp. 71–2.

Book Three: The Bad Self

page

88 **'does not differ in any essentials':** *Pluscarden Abbey*, Dom Augustine Holmes OSB (Heritage House, 2004), p. 28.

91 **'Death is stationed':** *Rule of St Benedict*, 72.4.

91 **'Truly, we are forbidden to do our own will':** *Rule of St Benedict*, 7.19.

91 **'I am truly a worm, not a man':** *Rule of St Benedict*, 7.51–52.

91 **I went on to read that 'the main essential' of Christian life:** *Our Purpose and Method*, Abbot Aelred Carlyle OSB (Pluscarden Abbey, 1907), p. 3.

93 **Food was often scarce . . . etc.:** *Medieval Britain*, John Gillingham and Ralph A. Griffiths (Oxford University Press, 2000), p. 101.

93 **Up to a tenth of the population:** *Medieval Britain*, John Gillingham and Ralph A. Griffiths (Oxford University Press, 2000), p. 69.

93 **with the majority of the rest:** *Medieval Britain*, John Gillingham and Ralph A. Griffiths (Oxford University Press, 2000), p. 75.

94 **Joseph Henrich writes that the 'cultural learning':** *The Secret of our Success*, Joseph Henrich (Princeton University Press, 2016), p. 36.

94 **A basic cue we look for is 'self-similarity':** *The Secret of our Success*, Joseph Henrich (Princeton University Press, 2016), p. 44.

94 **Another cue is age:** *The Secret of our Success*, Joseph Henrich (Princeton University Press, 2016), p. 46.

94 **Physical dominance is a cue that:** *The Secret of our Success*, Joseph Henrich (Princeton University Press, 2016), p. 130.

94 **These cues are success and prestige:** *The Secret of our Success*, Joseph Henrich (Princeton University Press, 2016), p. 37.

95 **Research suggests that we start mimicking people:** *The Secret of our Success*, Joseph Henrich (Princeton University Press, 2016), p. 42.

95 **these 'skill cues' begin to take on a more symbolic form, as 'success cues' . . . etc.:** *The Secret of our Success*, Joseph Henrich (Princeton University Press, 2016), pp. 38–40.

95 **we tend to look at who *other people* . . . etc.:** *The Secret of our Success*, Joseph Henrich (Princeton University Press, 2016), p. 42–4.

95 **'Once people have identified a person as worthy':** *The Secret of our Success*, Joseph Henrich (Princeton University Press, 2016), p. 42.

95 **They might give them gifts . . . etc.:** *The Secret of our Success*, Joseph Henrich (Princeton University Press, 2016), p. 119.

96 **they'll often mimic:** *The Secret of our Success*, Joseph Henrich (Princeton University Press, 2016), p. 42.

96 **this hum is actually 'an unconscious social instrument':** *Our Inner Ape*, Frans de Waal (Granta, 2005), p. 56.

96 **everyone else adjusts theirs to match it:** *Our Inner Ape*, Frans de Waal (Granta, 2005), p. 56; *The Secret of our Success*, Joseph Henrich (Princeton University Press, 2016), p. 56.

96 **'The visual attention of the bystanders provided a "prestige cue"':** *The Secret of our Success*, Joseph Henrich (Princeton University Press, 2016), p. 43.

96 **A clever study by a team including Henrich had pre-schoolers:** 'Prestige-biased cultural learning: bystander's differential attention to potential models influences children's learning', Maciej Chudek et al., *Evolution and Human Behavior* (January 2012), **33**(1), pp. 46–56.

96–7 **one of the stranger facets of modern celebrity culture . . . the 'Paris Hilton effect':** *The Secret of our Success*, Joseph Henrich (Princeton University Press, 2016), pp. 125–6.

97 **we're naturally triggered into this behaviour:** *The Secret of our Success*, Joseph Henrich (Princeton University Press, 2016), p. 43.

97 **we'll often internalize the things they've taught us:** *The Secret of our Success*, Joseph Henrich (Princeton University Press, 2016), p. 154.

98 **'A protagonist is a wilful character':** *Story*, Robert McKee (Methuen, 1999), p. 137.

99 **Psychologists describe the 'effectance motive':** *The Happiness Hypothesis*, Jonathan Haidt (Heinemann, 2006), p. 220.

99 **When people are left floating in darkened salt-water tanks:** *Brain and Culture*, Bruce Wexler (MIT Press, 2008), p. 76.

99 **409 people stripped of their phones and left alone in a room:** 'Just think: The challenges of the disengaged mind', Timothy D. Wilson et al., *Science* (July 2014), **345**(6192), pp. 75–7.

99 **The neurobiologist Robert Sapolsky has argued:** Video lecture: 'Dopamine Jackpot! Sapolsky on the Science of Pleasure', http://www.dailymotion.com/video/xh6ceu_dopamine-jackpot-sapolsky-on-the-science-of-pleasure_news.

99 **work by geneticist Professor Steve Cole:** I wrote about Steve Cole's work in the *New Yorker* ('A Better Kind of Happiness', *New Yorker*, 7 July 2016). Cole and his colleagues' work on eudaemonic happiness has received some passionate and sustained criticism from a group of researchers who seem to object, in part, to its apparent endorsement of a finding that has its roots in positive psychology. They have a history of repeated attempts at 'debunking' this research, Cole would then address their criticisms only for

them to then seek out new complaints. At the time of writing, this process was still ongoing.

Whilst I share these researchers' instinctive suspicion of positive psychology, it did seem to me that their particular pattern of behaviour perhaps betrayed an ideological, rather than a rational, motivation. This is, of course, only my opinion. I should also add that, just because someone is ideologically motivated, it doesn't automatically follow that they are wrong. As I suggest in the main text, this is early work, and should be treated as such. If you're interested in tracking the debate further, a good place to start is the excellent Jo Marchant's *Nature* report on the earlier trial, 'Immunology: The pursuit of happiness', *Nature* (November 2013), 503(7477), pp. 458–60.

100 **Further studies have found that people with a greater sense of purpose:** 'Purpose in Life as a Predictor of Mortality Across Adulthood', Patrick Hill and Nicholas Turiano, *Psychological Science* (May 2014), 25(7), pp. 1487–96.

100 **many decades of research by the psychologist Professor Brian Little . . . fifteen:** Interview with Brian Little and *Me, Myself and Us*, Brian Little (PublicAffairs, 2014), p. 183.

101 **observation about 'materialists' who don't believe in God:** *Our Purpose and Method*, Abbot Aelred Carlyle OSB (Pluscarden Abbey, 1907), p. 6.

106 **the monasteries were pioneers of an embryonic form of capitalism:** *The Victory of Reason*, Rodney Stark (Random House, 2005), pp. 55–61.

107 **As the social scientist Professor Rodney Stark has observed, where the Qur'an:** *The Victory of Reason*, Rodney Stark (Random House, 2005), p. 9.

107 **In the fifth century, St Augustine wrote that Christians should 'approach together':** *The Victory of Reason*, Rodney Stark (Random House, 2005), p. 11.

107 **Writes Stark, 'From the early days, the Church fathers':** *The Victory of Reason*, Rodney Stark (Random House, 2005), p. x.

107 **This is why Islam and Judaism are known as 'orthoprax':** *The Victory of Reason*, Rodney Stark (Random House, 2005), p. 8.

107 **according to Stark, was always 'on discovering God's nature':** *The Victory of Reason*, Rodney Stark (Random House, 2005), p. 5.

108 **Aristotle in the Lyceum:** Note from my excellent copy editor, Nicholas Blake. I had Aristotle teaching in the market square.

109 **One of the strangest chain of events:** My account of the childhood of Sigmund Freud was sourced from the following books: *Freud: Darkness in the Midst of Vision*, Louis Breger (John Wiley and Sons, 2000); *A Compulsion for Antiquity: Freud and the Ancient World*, Richard Armstrong (Cornell University Press, 2005); *Freud and Oedipus*, Peter L. Rudnytsky (Columbia University Press, 1987); *Classical Mythology*, Helen Morales (Oxford University Press, 2012).

109 **would help him fall asleep, it is thought, by quietly stroking his cock:** There's a bit of supposition here, but there is little doubt that Freud was sexually molested by Monica. He wrote that 'she was my instructress in sexual matters' and, elsewhere, that, 'It is well known unscrupulous nurses put crying children to sleep by stroking their genitals.' *Freud and Oedipus*, Peter L. Rudnytsky (Columbia University Press, 1987), p. 58.

110 **In 1873, as part of his final school examinations:** http://www.freud.org.uk/education/timeline/.

111 **As historian Professor Peter Rudnytsky has observed, 'The coincidence between his biographical accidents of birth and the Oedipus . . .':** *Freud and Oedipus*, Peter L. Rudnytsky (Columbia University Press, 1987), p. 16.

111 **'Every new arrival on this planet is faced by the task of . . .':** Quoted in *Freud and Oedipus*, Peter L. Rudnytsky (Columbia University Press, 1987), pp. 51–2.

112 **It's well known by modern psychologists that:** For much more on this, see *Mindwise*, Nicholas Epley (Penguin, 2014).

112 **Epley, who's studied this effect, writes that, 'Brown-bread lovers':** *Mindwise*, Nicholas Epley (Penguin, 2014), p. 101.

112 **When participants in a brain scanner were interrogated about God's views:** 'Believers' estimates of God's beliefs are more egocentric than estimates of other people's beliefs', Nicholas Epley et al., *Proceedings of the National Academy of Sciences of the United States of America*, vol. 106, no. 51.

112 **people's own views have been seen to change in concert:** 'Believers' estimates of God's beliefs are more egocentric than estimates of other people's beliefs', Nicholas Epley et al., *Proceedings of the National Academy of Sciences of the United States of America* (December 2009), **106**(51), pp. 21533–8.

113 **'When others' minds are unknown,' writes Epley:** *Mindwise*, Nicholas Epley (Penguin, 2014), p. 111.

113 **'Freud saw the analyst as an Oedipus figure':** *Classical Mythology*, Helen Morales (Oxford University Press, 2012), p. 71.

113 **'To be sure, psychoanalysis was born, in part':** *Classical Mythology*, Helen Morales (Oxford University Press, 2012), p. 74.

113 **Without the myths of Ancient Greece, she suggests:** *Classical Mythology*, Helen Morales (Oxford University Press, 2012), p. 69.

113 **It's not a coincidence, of course, that this renaissance was centred . . . etc.:** *The Renaissance*, Jerry Brotton (Oxford University Press, 2006), pp. 24–8.

114 **In 1936, in an incident that surely seemed to him of little consequence:** This account is taken from Fritz Perls' 1969 autobiography, *In and Out the Garbage Pail* (Bantam, 1972). Perls declined to number the pages. Additional contextual details were sourced from *Esalen: America and the Religion of No Religion*, Jeffrey J. Kripal (University of Chicago Press, 2007) and *The Upstart Spring: Esalen and the Human Potential Movement, The First Twenty Years*, Walter Truett Anderson (iUniverse, 2004).

Book Four: The Good Self

page

121 **from our fetishization of personal authenticity and 'being real':** Of course, Esalen isn't the sole source of these ideas. Authenticity as a pronounced value is common in many working-class cultures, for example. One of the interesting things about Esalen is that it managed to repackage and redefine many ideas for the middle classes.

121 **lived under 'a system of socially':** *Smile or Die*, Barbara Ehrenreich (Granta, 2010), p. 75.

121 **'took for granted the reality':** *God's Salesman: Norman Vincent Peale and the Power of Positive Thinking*, Carol V. R. George (Oxford University Press, 1993), p. viii.

123 **'to stimulate youths to apply themselves diligently':** *Self Help*, Samuel Smiles (John Murray, 1859), p. 9 (1996 edition).

124 **expat Yorkshire plumber named Smith Wigglesworth:** *Stories of the Supernatural: Finding God in Walmart and Other Unlikely Places*, Tyler Johnson (Destiny Image, 2010), p. 34.

124 **James defined mind-cure as the 'intuitive belief in the all-saving power':** *The Varieties of Religious Experience*, William James (Longmans, Green, 1902), p. 93.

124 **Mind-cure's forefather was a clockmaker from New England**

named Phineas Quimby: *One Simple Idea: How Positive Thinking Reshaped Modern Life*, Mitch Horowitz (Crown, 2014), p. 126.

124 **'Quimby's method was to sympathetically':** *One Simple Idea: How Positive Thinking Reshaped Modern Life*, Mitch Horowitz (Crown, 2014), p. 126.

124 **In 1862 Quimby treated Mary Baker Eddy:** *One Simple Idea: How Positive Thinking Reshaped Modern Life*, Mitch Horowitz (Crown, 2014), p. 132.

125 **economists sometimes call the 'Great Compression . . . between 1945 and 1975:** *The Rise and Fall of American Growth: The US Standard of Living Since the Civil War*, Robert J. Gordon (Princeton University Press, 2016), p. 613.

125 **The GI Bill:** *The Rise and Fall of American Growth: The US Standard of Living Since the Civil War*, Robert J. Gordon (Princeton University Press, 2016), p. 606.

125 **between 1929 and 1945, lower incomes would grow faster:** *The Rise and Fall of American Growth: The US Standard of Living Since the Civil War*, Robert J. Gordon (Princeton University Press, 2016), p. 613.

125 **'a golden age for millions of high school graduates', writes the economist Professor Robert Gordon, 'who without a college education . . .':** *The Rise and Fall of American Growth: The US Standard of Living Since the Civil War*, Robert J. Gordon (Princeton University Press, 2016), p. 609.

126 **It was around this time, as Susan Cain has famously documented, that the provost of Harvard University:** *Quiet*, Susan Cain (Penguin, 2013), p. 127.

126 **'If we are our own chief problem, the basic reason must be found in the type of thoughts':** Quoted in *Smile or Die*, Barbara Ehrenreich (Granta, 2010), p. 92.

126 **As sociologist Professor John Hewitt so adroitly puts it, 'The era of "character" vanished:'** *Dilemmas of the American Self*, John P. Hewitt (Temple, 1989), p. 96.

127 **It was feared the future would be a 'technocracy' in which freedom:** *From Counterculture to Cyberculture*, Fred Turner (University of Chicago Press, 2006), p. 29.

127 **wrote Lewis Mumford in his 1967 book, *The Myth of the Machine*:** Quoted in *From Counterculture to Cyberculture*, Fred Turner (University of Chicago Press, 2006), p. 29.

128 **a five-month trip to China as a theology student:** *Carl Rogers:*

The Quiet Revolutionary, Carl R. Rogers and David E. Russell (Penmarin, 2002), p. 62.

129 **'the innermost core of man's nature'**: Quoted in *The Therapeutic State: Justifying Government at Century's End*, James L. Nolan Jnr (New York University Press, 1998), Kindle location 191.

129 **As an eighty-three-year-old, in 1985, Carl Rogers was still complaining:** *Carl Rogers: The Quiet Revolutionary*, Carl R. Rogers and David E. Russell (Penmarin, 2002), p. 244.

129 **'If one needs any evidence'**: *Carl Rogers: The Quiet Revolutionary*, Carl R. Rogers and David E. Russell (Penmarin, 2002), p. 244.

129 **'We had the idea that if it was good for neurotics, it would be good for normals . . . There were some 615 nuns when we began . . . etc.'**: Interview accessed at: https://www.ewtn.com/library/PRIESTS/COULSON.TXT.

130 **Aldous Huxley, gave a lecture at the University of California, Berkeley:** *The Upstart Spring: Esalen and the Human Potential Movement, The First Twenty Years*, Walter Truett Anderson (iUniverse, 2004), pp. 10–11.

131–2 Unless specifically noted, my account of the story of Esalen was sourced from the two major publicly available histories, both of which are superb and well worth reading on their own terms. They are: *The Upstart Spring: Esalen and the Human Potential Movement, The First Twenty Years*, Walter Truett Anderson (iUniverse, 2004) and *Esalen: America and the Religion of No Religion*, Jeffrey J. Kripal (University of Chicago Press, 2007).

131 **Sociologist Professor Marion Goldman writes that one of the reasons Esalen attracted thousands of Americans:** *The American Soul Rush: Esalen and the Rise of Spiritual Privilege*, Marion Goldman (New York University Press, 2012), p. 4.

132 **'Esalen played a critical role in introducing'**: *The American Soul Rush: Esalen and the Rise of Spiritual Privilege*, Marion Goldman (New York University Press, 2012), p. 1.

132 **'Millions of contemporary Americans identify themselves as spiritual'**: *The American Soul Rush: Esalen and the Rise of Spiritual Privilege*, Marion Goldman (New York University Press, 2012), p. 1.

132 **'The basic assumption that God is part of all beings'**: *The American Soul Rush: Esalen and the Rise of Spiritual Privilege*, Marion Goldman (New York University Press, 2012), p. 12.

133 **'Confidentiality. Feel free to share what you see in here, but don't attach names'**: All names in my account of my experience at Esalen have been altered, except that of Paula Shaw.

136–7 **'a painful and in some ways historically significant meeting':** *The Upstart Spring: Esalen and the Human Potential Movement, The First Twenty Years,* Walter Truett Anderson (iUniverse, 2004), p. 93.

137 **stroking the genitals of any he found acquiescent:** *The American Soul Rush: Esalen and the Rise of Spiritual Privilege,* Marion Goldman (New York University Press, 2012), p. 36.

137 **He took to wearing slippers:** *The American Soul Rush: Esalen and the Rise of Spiritual Privilege,* Marion Goldman (New York University Press, 2012), p. 127.

137–8 **His groups operated 'Strictly on the "I and thou, here and now" basis,' . . . 'change paper people into real people':** See 'Fritz Perls A Session with College Students', https://www.youtube.com/watch?v=ZsZqJXf4vMI, 'Fritz Perls Here and Now . . .', https://www.youtube.com/watch?v=6AAgeT1X5oI.

138 **was accused by Fritz of 'absolute phoniness':** *The American Soul Rush: Esalen and the Rise of Spiritual Privilege,* Marion Goldman (New York University Press, 2012), p. 129.

139 **'doctors, social workers, clinical psychologists, teachers, students, business executives, engineers, housewives', according to a contemporary *New York Times* account:** 'Joy is the Prize: A Trip to Esalen Institute', Leo E. Litwak, *New York Times,* 31 December 1967.

139 **group sex:** *The American Soul Rush: Esalen and the Rise of Spiritual Privilege,* Marion Goldman (New York University Press, 2012), p. 13.

139 **young women in sheer gowns:** 'Joy is the Prize: A Trip to Esalen Institute', Leo E. Litwak, *New York Times,* 31 December 1967.

139 **had an affair with co-founder Richard Price:** 'Jane Fonda: "I never was a hippie!"', Andrew O'Hehir, *Salon,* 9 June 2012.

139 **a state Schutz described as 'the feeling that comes from the fulfilment of one's potential':** *The Upstart Spring: Esalen and the Human Potential Movement, The First Twenty Years,* Walter Truett Anderson (iUniverse, 2004), p. 157.

140 **'There is no such thing as a victim of circumstance,' he said:** Interview, *San Francisco Chronicle,* 23 October 1978.

140 **'Under this Encounter Contract I say':** *Esalen: America and the Religion of No Religion,* Jeffrey J. Kripal (University of Chicago Press, 2007), p. 168.

141 **'Like Fritz,' writes Anderson, 'Schutz made the taking of responsibility':** *The Upstart Spring: Esalen and the Human Potential Movement, The First Twenty Years,* Walter Truett Anderson (iUniverse, 2004), p. 178.

141 **a guest on *The Tonight Show* starring Johnny Carson:** *The American*

Soul Rush: Esalen and the Rise of Spiritual Privilege, Marion Goldman (New York University Press, 2012), p. 6.

142 **as the philosopher Julian Baggini has observed, 'our minds are just one'**: *The Ego Trick*, Julian Baggini (Granta, 2011), p. 119.

142–3 **In his book *The Self Illusion*, Bruce cites its creator, the sociologist Charles Horton Cooley:** *The Self Illusion*, Bruce Hood (Constable, 2011), p. 51.

144 **Jean-Paul Sartre, who noticed it in a waiter . . . etc.:** *Being and Nothingness*, Jean-Paul Sartre (Gallimard, 1943), p. 59.

145 **This is likely because Mum and Dad are treating us as the person:** 'Stop Reverting to Childhood on Your Holiday Visit Home', Melissa Dahl, *New York Magazine*, 25 November 2015.

145 **A team led by psychologist Professor Mark Snyder examined some of these effects:** 'Social Perception and Interpersonal Behavior: On the Self-Fulfilling Nature of Social Stereotypes', Mark Snyder et al., *Journal of Personality and Social Psychology* (1977), **35**(9), pp. 655–66.

146 **Other researchers have found that:** 'Behavioral Confirmation of the Loneliness Stereotype', Ken J. Rotenberg, Jamie A. Gruman and Mellisa Ariganello, *Basic and Applied Social Psychology* (2002), **24**(2), pp. 81–9.

146 **it's the environments that really do the switching:** *The Self Illusion*, Bruce Hood (Constable, 2011), p. x.

146 **The psychologists Dan Ariely and George Loewenstein explored our multiple nature:** 'The Heat of the Moment: The Effect of Sexual Arousal on Sexual Decision Making', Dan Ariely and George Loewenstein, *Journal of Behavioral Decision Making* (April 2006), **19**(2), pp. 87–98. See also commentary in *Predictably Irrational*, Dan Ariely (HarperCollins, 2008), p. 96.

147 **David Eagleman writes that our brains are 'built of multiple overlapping experts . . . etc':** *Incognito*, David Eagleman (Canongate, 2012), p. 108.

158 **There were nearly one hundred unaffiliated so-called 'Little Esalens':** *The American Soul Rush: Esalen and the Rise of Spiritual Privilege*, Marion Goldman (New York University Press, 2012), p. 5.

157 **the Institute's work was being seriously discussed:** *The American Soul Rush: Esalen and the Rise of Spiritual Privilege*, Marion Goldman (New York University Press, 2012), p. 6.

158 **Thousands of the nation's psychiatrists:** 'Joy is the Prize: A Trip to Esalen Institute', Leo E. Litwak, *New York Times*, 31 December 1967.

158 **ten thousand people in its first two months:** 'Joy is the Prize: A Trip to Esalen Institute', Leo E. Litwak, *New York Times*, 31 December 1967.

158 **founders Murphy and Price began downplaying:** *The American Soul Rush: Esalen and the Rise of Spiritual Privilege*, Marion Goldman (New York University Press, 2012), p. 38.

158 **'Many reasons were given for ending it,' writes Anderson:** *The Upstart Spring: Esalen and the Human Potential Movement, The First Twenty Years*, Walter Truett Anderson (iUniverse, 2004), p. 234.

161 **Today, some researchers believe:** See *Loneliness: Human Nature and the Need for Social Connection*, John Cacioppo and William Patrick (WM Norton, 2009).

162 **Professor James Coan at the University of Virginia has suggested that we only:** 'Familiarity promotes the blurring of self and other in the neural representation of threat', Lane Beckes et al., *Social Cognitive and Affective Neuroscience* (August 2013), 8(6), pp. 670–7.

162 **Other studies, by researchers at China's Shenzhen University:** 'Social hierarchy modulates neural responses of empathy for pain', Chunliang Feng et al., *Social Cognitive and Affective Neuroscience* (March 2016), 11(3), pp. 485–95.

163 **psychologist Professor Jonathan Haidt calls us '10 per cent bee':** *The Righteous Mind*, Jonathan Haidt (Allen Lane, 2012), p. xv.

166 **In the months leading up to his death, in 1970, the humanistic psychologist Abraham Maslow:** La Jolla Program Newsletter, Self-Esteem Task Force Edition, vol. XX, no. 8, April 1988, pp. 1–2.

166 **'After several years in California, Carl got so tired':** Letter from W. K. Coulson to Robert Ball, Exec Director CA Task Force etc., 22 March 1988, Inventory of the Task Force to Promote Self-esteem and Personal and Social Responsibility Records, Sacramento, California.

166 **In December 1973:** The conference marked a disillusionment in some of the Eastern philosophies that were often founded in Confucianism. 'The human potential movement was beginning to realize that the mystical was not the ethical,' writes Kripal (*Esalen: America and the Religion of No Religion*, University of Chicago Press, 2007, p. 288), 'that spiritual authority is often more or less identical to moral tyranny, and that the Asian systems, embedded as they are in ancient hierarchical systems, can only deliver so much to a modern liberal democracy.'

167 **Erhard, writes Truett, 'Americanized the human potential':** *The*

Upstart Spring: Esalen and the Human Potential Movement, The First Twenty Years, Walter Truett Anderson (iUniverse, 2004), p. 254.

166 **fifty thousand people took the workshop in its first four years:** 'The New Narcissism', Peter Marin, *Harper's Magazine*, October 1975.

167 **It represented a kind of business-ification of Human Potential methods:** *The Upstart Spring: Esalen and the Human Potential Movement, The First Twenty Years*, Walter Truett Anderson (iUniverse, 2004), pp. 262–3.

167 **'He got a million dollars' worth of advice from some':** *The Upstart Spring: Esalen and the Human Potential Movement, The First Twenty Years*, Walter Truett Anderson (iUniverse, 2004), p. 148.

169 **having first visited in 1962:** 'Esteeming to retirement', Timothy Roberts, *Silicon Valley Business Journal*, 10 February 2002.

169 **that all men somehow possess a divine potentiality:** 'Joy is the Prize: A Trip to Esalen Institute', Leo E. Litwak, *New York Times*, 31 December 1967.

Book Five: The Special Self

page

173 **So there was this girl, name of Alyssa Rosenbaum:** My telling of the Ayn Rand, Nathaniel Branden and Alan Greenspan story is sourced from the following accounts, except where specifically noted: *Ayn Rand and the World She Made*, Anne C. Heller (Anchor, 2009); *Ayn Rand Nation*, Gary Weiss (St Martin's Griffin, 2012); *My Years with Ayn Rand*, Nathaniel Branden (Jossey-Bass, 1999); *Maestro: Greenspan's Fed and the American Boom*, Bob Woodward (Simon and Schuster, 2000); *Alan Shrugged: Alan Greenspan, the World's Most Powerful Banker*, Jerome Tuccille (John Wiley, 2002); *Alan Greenspan: The Oracle Behind the Curtain*, E. Ray Canterbury (World Scientific, 2006).

173 **They confiscated her father's chemist shop:** *The Life of I*, Anne Manne (Melbourne University Press, 2014), Kindle location 2599.

173 **'I thought, right then, that this idea was evil':** *My Years with Ayn Rand*, Nathaniel Branden (Jossey-Bass, 1999), pp. 62–3.

173 **'Men have been taught that the highest virtue is not to achieve':** *The Fountainhead*, Ayn Rand (Bobbs-Merrill, 1943), p 513.

174 **In reality, the Collective was a kind of cult . . . etc.:** *My Years with Ayn Rand*, Nathaniel Branden (Jossey-Bass, 1999), p. 226.

174 **'We thought of ourselves as being the instigators of a revolution':**

All Watched Over, Machines of Loving Grace, episode one (BBC, 2011).

174 **Rand claimed that the only thinker to have ever influenced her was the father of individualism:** Rand on *The Mike Wallace Interview*, available at https://www.youtube.com/watch?v=HKdoToQDooo.

174 **'Man is entitled to his own happiness and he must achieve it himself . . .':** Rand on *The Mike Wallace Interview*, available at https://www.youtube.com/watch?v=HKdoToQDooo.

175 **'The nature of his self-evaluation has profound effects . . .':** *The Psychology of Self-Esteem*, Nathaniel Branden (Jossey-Bass, 1969), p. 109.

175 **became known as 'the father of the self-esteem movement':** *Ayn Rand and the World She Made*, Anne C. Heller (Anchor, 2009), p. 378.

176 **'a free mind and a free market are corollaries':** *Capitalism: The Unknown Ideal*, Ayn Rand (New American Library, 1966), p. 47.

176 **gave him the nickname the Undertaker:** *My Years with Ayn Rand*, Nathaniel Branden (Jossey-Bass, 1999), p. 113.

176 **'He came alive with an excitement':** *My Years with Ayn Rand*, Nathaniel Branden (Jossey-Bass, 1999), p. 113.

176 **Greenspan thought her ideas 'radiantly exact':** *Ayn Rand and the World She Made*, Anne C. Heller (Anchor, 2009), p. 275.

176 **'What she did,' he said, 'was to make me see that capitalism . . .':** *Alan Greenspan: The Oracle Behind the Curtain*, E. Ray Canterbury (World Scientific, 2006), p. 6.

176 **'their values, how they work, what they do':** *Ayn Rand and the World She Made*, Anne C. Heller (Anchor, 2009), p. 398.

176 **she changed his nickname from the Undertaker to the Sleeping Giant:** *Ayn Rand and the World She Made*, Anne C. Heller (Anchor, 2009), p. 242.

176–7 **'It is precisely the "greed" of the businessman':** *Ayn Rand and the World She Made*, Anne C. Heller (Anchor, 2009), p. 276.

177 **Racism, she believed, was 'the lowest, most crudely':** *The Virtue of Selfishness*, Ayn Rand (New American Library, 1964), p. 126.

177 **over the next two years, she'd have sex:** *My Years with Ayn Rand*, Nathaniel Branden (Jossey-Bass, 1999), p. 219.

177 **She reacted with vengeful rage, hitting him three:** *Ayn Rand and the World She Made*, Anne C. Heller (Anchor, 2009), p. 372.

177 **'You have rejected me?':** *Ayn Rand and the World She Made*, Anne C. Heller (Anchor, 2009), p. 370.

178 **At one point word reached Branden that the Collective:** 'Looking

Back Objectively', Trish Todd, *Publisher's Weekly*, 10 January 1986, p. 56.

178 **had been high enough to buy what the nation itself had been making:** *Neoliberalism*, Manfred B. Steger and Ravi K. Roy (Oxford University Press, 2010), p. 7.

178 **after insistent urging by Rand:** *Alan Greenspan: The Oracle Behind the Curtain*, E. Ray Canterbury (World Scientific, 2006), p. 7.

178 **'neoliberalism':** My account of the rise of neoliberalism was sourced from: *Masters of the Universe: Hayek, Friedman and the birth of Neoliberal Politics*, Daniel Stedman Jones (Princeton University Press, 2012); *The Constitution of Liberty*, F. A. Hayek (Routledge, 1960); *The Rise and Fall of American Growth: The US Standard of Living Since the Civil War*, Robert J. Gordon (Princeton University Press, 2016); *Neoliberalism: A Brief History*, David Harvey (Oxford University Press, 2005); *Neoliberalism*, Manfred B. Steger and Ravi K. Roy (Oxford University Press, 2010); *Ill Fares the Land*, Tony Judt (Penguin, 2010); *Globalisation*, Manfred B. Steger (Oxford University Press, 2013).

179 **'There is more than a superficial similarity between':** As quoted in *Masters of the Universe: Hayek, Friedman and the birth of Neoliberal Politics*, Daniel Stedman Jones (Princeton University Press, 2012), p. 30.

179 **'Mr Fluctooations'** Stephen Metcalf, 'Neoliberalism: the idea that swallowed the world', *Guardian*, 18 August 2017.

179 **as being a betrayal of their Ancient Greek inheritance:** *Masters of the Universe: Hayek, Friedman and the birth of Neoliberal Politics*, Daniel Stedman Jones (Princeton University Press, 2012), p. 59.

179 **'Economic control is not merely control of a sector of human life':** As quoted in *Masters of the Universe: Hayek, Friedman and the birth of Neoliberal Politics*, Daniel Stedman Jones (Princeton University Press, 2012), p. 69.

179 **'coercion of some by others is reduced as much as possible':** *The Constitution of Liberty*, F. A. Hayek (Routledge, 1960), p. 11.

180 **The human world would become a kind of game:** I first read about neoliberalism turning society into a game in Will Davies' essay 'How "competitiveness" became one of the great unquestioned virtues of contemporary culture', which was posted on the LSE website on 19 May 2014. http://blogs.lse.ac.uk/politicsandpolicy/the-cult-of-competitiveness/.

180 **they'd 'perform a necessary service' by 'experimenting with new styles of living':** *The Constitution of Liberty*, F. A. Hayek (Routledge,

1960), p. 41. I first read about this idea of Hayek's in George Monbiot's column 'Neoliberalism: the deep story that lies beneath Donald Trump's triumph', published 14 November 2016 in the *Guardian.*

180 **at a meeting** Alan O. Ebenstein, *Friedrich Hayek: A Biography* (St Martin's Press, 2014), p. 291.

180 **'the most powerful critique** Alan O. Ebenstein, *Friedrich Hayek: A Biography* (St Martin's Press, 2014), p. 291.

180 **'There is no such thing as society. There are only individual men and women and families':** Quoted in *Neoliberalism: A Brief History*, David Harvey (Oxford University Press, 2005), p. 23.

181 **made Greenspan 'the single most powerful figure':** *Alan Greenspan: The Oracle Behind the Curtain*, E. Ray Canterbury (World Scientific, 2006), p. 1.

181 **the 'Central Banker of Neoliberalism':** 'The central banker of neoliberalism: Alan Greenspan steps down as Fed Chief', Joel Geier, http://socialistworker.org/2006-1/575/575_06_Greenspan.shtml.

181 **act as both groundskeeper and referee:** *Neoliberalism*, Manfred B. Steger and Ravi K. Roy (Oxford University Press, 2010), p. 2.

181 **There would be no more 'citizens', but 'clients' or 'customers':** *Neoliberalism*, Manfred B. Steger and Ravi K. Roy (Oxford University Press, 2010), p. 113.

182 **The World Bank and the IMF would assist in this mission:** *Globalisation*, Manfred B. Steger (Oxford University Press, 2013), p. 57.

182 **they'd eventually rival nation states:** *Globalisation*, Manfred B. Steger (Oxford University Press, 2013), p. 54.

182 **'Economics are the method,' said Thatcher:** Quoted in *Neoliberalism: A Brief History*, David Harvey (Oxford University Press, 2005), p. 23.

183 **As a study of over 300 million births starting in 1880:** 'Fitting In or Standing Out: Trends in American Parents' Choices for Children's Names, 1880–2007', Jean M. Twenge et al., *Social Psychological and Personality Science* (2010), 1(1), pp. 19–25.

183 **Vasco was an intense, brooding, scruffy man:** My account of the life and work of John Vasconcellos was sourced from documentary research carried out at the state archives in Sacramento, where the records from the self-esteem task force are kept, and the archives at the University of California, Santa Barbara, where Vasconcellos' personal records are kept. As well as this, I used contemporary

news reports, interviews and profiles. I am grateful to Vasconcellos' friend David E. Russell from the UCSB Library Oral History Program for providing me with the transcript of his 1993 interview with Vasconcellos' mother. I supplemented these sources with my own interviews with people who knew and worked with Vasconcellos.

183 **'thought people couldn't take care of themselves' . . . etc.:** 'The West Interview: John Vasconcellos', *Mercury News*, 4 January 1987.

183 **Declaring his belief in 'individuality, freedom':** 'The Self-Esteem Task Force: Making California Feel Good', Siobhan Ryan, *California Journal*, June 1990.

183 **he praised conservatives for being 'big on economic freedom':** 'Human potential and politics', Bea Pixa, *San Francisco Examiner*, 4 October 1979.

184 **a stick man with 'I' for individual:** Profile, Richard Trainer, *California Magazine*, October 1987.

184 **deeply influenced by his experiences at Esalen:** Interview, Andrew Mecca.

184 **more than ten billion dollars:** 'Pondering Self-Esteem', David Gelman with George Raine, *Newsweek*, 2 March 1987.

184 **Until the 1980s, self-esteem had been of interest mainly:** A 1987 report on the task force noted that their assertion that 'self-esteem is deemed laughable today . . . only because it has until now been largely ignored.' 'Self-esteem panel says it's no joking matter', Elizabeth Fernandez, *San Francisco Examiner*, 13 February 1987.

186 **'I lost by one vote. Mine,' he said . . . etc.:** Speech, 7th Annual Conference on Self-Esteem.

186 **'I found myself and my identity and my life . . . For the next year, at least weekly':** Carl Rogers eulogy, 21 February 1987.

187 **participated in a series of eight workshops:** Preliminary Guide to the John Vasconcellos Papers, Online Archive of California, available at http://www.oac.cdlib.org/findaid/ark:/13030/kt3c601926/entire_text/.

187 **'almost my second father':** Carl Rogers eulogy, 21 February 1987.

187 **thought the mass should never have been translated:** 'The Case of the Liberated Legislator', Ralph Keyes, *Human Behaviour*, October 1974.

187 **'a cross between a rock star and a drug smuggler':** Profile, Richard Trainer, *California Magazine*, October 1987.

188 **'I've seen John so mad he'd pretty nearly froth':** 'The Case of the

Liberated Legislator', Ralph Keyes, *Human Behaviour*, October 1974.

188 **'It's all life and death with him . . .':** 'The "Touchy Feely" Legislator', Gail Schontzler, *California Journal*, December 1975.

188 **He'd become particularly enraged:** 'The Case of the Liberated Legislator', Ralph Keyes, *Human Behaviour*, October 1974.

188 **to hold his hand:** 'The Unsettled Self-Esteem of John Vasconcellos', Jacques Leslie, *Los Angeles Times*, 23 August 1987.

188 **'It's fairly easy for you to justify':** Memo from 'Sherry' dated 1974, The John Vasconcellos Papers, Department of Special Collections, Davidson Library, University of California, Santa Barbara.

188 **last for days, sometimes months:** 'Bachelor? But don't preface it with "confirmed"', Carol Sarasohn, *San Jose Mercury*, 16 August 1978.

188 **'Our culture is dying,' he wrote:** 'A New Vision of Man, Human Nature, and Human Potential', John Vasconcellos, *Humanist*, November/December 1972.

189 **'the unlikely marriage of Esalen and Sacramento':** 'The "Touchy Feely" Legislator', Gail Schontzler, *California Journal*, December 1975.

190 **a special form of 'gentle birthing':** Campaign document, undated.

190 **the idea that children having sex with their parents . . . etc.:** Vasconcellos kept a collection of cuttings that had this idea in common. One had, circled in red pen, the following paragraph: '"Patty's not the first woman to have incestuous relations with her father. Thousands of women do and always have." And now, he added, in an age when very little is considered taboo, more women even admit that they're actually proud of the fact that it was their fathers who taught them how to make love.'

190 **there was arguably only one person in the Capitol:** Some accounts have JV as the second most powerful person in the legislature, others place him behind the speaker.

191 **tiny brushes swimming through his arteries:** 'What's so bad about feeling good?', Jon Matthews, *Sacramento Bee*, 10 February 1987.

191 **'sit in committee and eat 1100 cookies':** 'The Unsettled Self-Esteem of John Vasconcellos', Jacques Leslie, *Los Angeles Times*, 23 August 1987.

191 **'You walk around here on eggshells':** 'The Unsettled Self-Esteem of John Vasconcellos', Jacques Leslie, *Los Angeles Times*, 23 August 1987.

191 **A gossipy item in the *Sacramento Bee*:** 'Self-esteem's prophet slips', Bee Capitol Bureau, *Sacramento Bee*, 5 July 1987.

192 **'So I got shrewd,' said Vasco:** Transcript of undated speech entitled 'Self Esteem Talk'.

193 **'I've never thought of it that way . . . etc.':** The bulk of this account was taken from JV's draft of his introduction to *The Social Importance of Self Esteem*.

193 **'Big John is very influential':** 'On Self Esteem', *San Francisco Chronicle*, 29 September 1986.

194 **'I've been willing to break new ground . . . etc.':** 'What's so bad about feeling good?', Jon Matthews, *Sacramento Bee*, 10 February 1987.

194 **Johnny Carson cracked wise:** 'When the Smug Laughs Stop', *Long Beach Press Telegram*, 19 February 1990.

194 **'You could buy the Bible for $2.50':** 'What's so bad about feeling good?', Jon Matthews, *Sacramento Bee*, 10 February 1987.

194 **The *San Francisco Examiner* called the idea 'laughable':** 'Self-esteem panel says it's no joking matter', Elizabeth Fernandez, *San Francisco Examiner*, 13 February 1987.

194 **The *Pittsburgh Post Despatch* wrote, 'California':** 'Self-esteem panel: another state joke?', *Pittsburgh Post Despatch*, 13 February 1987.

194 **'MAYBE FOLKS WOULD FEEL BETTER':** 'Maybe Folks Would Feel Better If They Got To Split the $735,000', Carrie Dolan, *Wall Street Journal*, 9 February 1987.

194 **Earlier, the *New York Times* had . . . .:** 'Now, The California Task Force to Promote Self-Esteem', *New York Times*, 11 October 1986.

194 **as one Los Angeles newspaper acknowledged:** 'Still Getting the Feel of Things', Deborah Hastings, *Herald Examiner* (date unclear in clipping, but apparently 11 September 1988).

194 **'I'm sick of it.' . . . etc.:** 'Assemblyman Dead Serious About Self-Esteem Panel', Robert B. Gunnion, *San Francisco Chronicle*, 9 February 1987.

195 **'I'm sick of the scornful stuff':** 'What's so bad about feeling good?', Jon Matthews, *Sacramento Bee*, 10 February 1987.

195 **'It's bizarre to me that someone':** 'Creator says task force gaining esteem', Jim Boren, *Fresno Bee*, 20 June 1987.

195 **His mother had been . . . etc.:** 'John Vasconcellos in conversation with Tricia Crane', *Los Angeles Herald Examiner*, 13 April 1987.

195 **'terrible, cynical, skeptical and cheap' . . . etc.:** 'Panel to Study Self-Esteem', *Education Week*, 5 November 1986.

195 **more than two thousand calls and letters:** 'Esteem: "Comic-strip"

task force members make their public debut', Jon Matthews, *Sacramento Bee*, 26 March 1987.

195 **almost four hundred applications:** Annual Progress Report, January 1988, p. 6.

196 **broke state records:** 'State Self-Esteem Committee is Getting a Lot More Respect', McClatchy News Service, *LA Daily Journal*, 23 February 1988.

196 **Fan-mail outnumbered:** 'Checking In With the State's Task Force, The Quest for Self-Esteem', Beth Ann Krier, *Los Angeles Times*, 14 June 1987.

196 **More than three hundred:** Annual Progress Report, January 1989, p. 19.

196 **a national figure:** 'Interview with John Vasconcellos', Tricia Crane, *Los Angeles Herald Examiner*, 13 April 1987.

196 **'I've gotten more attention':** 'The Unsettled Self-Esteem of John Vasconcellos', Jacques Leslie, *Los Angeles Times*, 23 August 1987.

196 **'The purpose of the task force was to':** 'Doonesbury vs. Self-Esteem', Arnold Hamilton, *Mercury News*, undated.

196 **female psychologist who argued:** 'Checking In With the State's Task Force, The Quest for Self-Esteem', Beth Ann Krier, *Los Angeles Times*, 14 June 1987.

196 **'even better than the stuff in Doonesbury':** 'Yeah, but tell me how you feel', John Corrigan, *Los Angeles Daily News*, 28 February 1988.

196 **Virginia Satir had asked her fellow members to close . . . etc.:** 'They feel better already', Bill Johnson, *LA Herald*, 2 February 1988.

198 **'made significant contributions to our task' . . . etc.:** Interview, Bob Ball.

198 **'encouraging the child to be in love with:** 'Notes on Self-Esteem for the California Task-Force', Nathaniel Branden.

198 **attacked the 'cascade of invective and ridicule':** Submission to the task-force: Self Respect, Not Arrogance.

198 **300 per cent:** 'When business boosts self-esteem', Robert J. McGarvey, *Kiwanis Magazine*, October 1989.

199 **This, it was agreed, was a significant coup:** *Esteem* newsletter, vol. 2, February 1988.

199 **'enormous credibility and a base to operate from':** 'Human frontier, self-esteem panel a "scouting party"', *Marin Independent Journal*, 26 March 1987.

199–200 **decision to publish one worth 'rejoicing':** Memo to task force members, 1 November 1988.

200 **it would be 'revolutionary research'**: *Esteem* newsletter, vol. 1, November 1987.
200 **of 'historic importance'**: Annual progress report, January 1988.
200 **'Through a contract negotiated'**: Second annual progress report, January 1989.
201 **glowing wire piece**: 'Self-esteem panel gaining some respect', *San Francisco Examiner*, 2 January 1989.
201 **'Self-Esteem panel finally being taken seriously'**: *Tribune*, Oakland, California, 1 January 1989.
201 **'Commission on Self-Esteem Finally Getting Some Respect'**: *Orange County Register*, 2 January 1989.
201 **'Boopsie Board Gains Respect'**: *North County Blade Tribune*, 1 January 1989.
201 **'I think we're now gaining a great deal of credibility'**: *Orange County Register*, 2 January 1989.
201 **The Duke was so impressed**: Letter, 1 October 1989.
201 **'There is magic in me because I believe in myself'**: 'Uplifting panel winds down', Jon Matthews, *Sacramento Bee*, 12 April 1989.
201 **Executive Director Bob Ball rightly called 'a master politician'**: Interview, Bob Ball.
202 **'a giant step for humankind' . . . etc.**: '"Self-esteem is the social vaccine" – Task force says if you feel good, you usually are good', Bill Ainsworth, *Sacramento Union*, 24 January 1990.
202 **Bill Clinton**: Letter from Clinton to JV in Task Force archive.
202 **Barbara Bush and Colin Powell**: *The Therapeutic State: Justifying Government at Century's End*, James L. Nolan Jnr (New York University Press, 1998), Kindle location 3059.
202 **'once the butt of "only-in-California" jokes'**: 'Self-esteem proposals ambitious', Steven A. Capps, *San Francisco Examiner*, 25 January 1990.
202 **The *Philadelphia Enquirer***: 'The Confidence Man', Carol Horner, 7 February 1990.
202 ***Time Magazine***: 'Learning Self Esteem', 29 January 1990.
202 **The *Ledger***: 'Task force finds good citizens are those with high self-esteem', Joan Morris, 22 January 1990.
202 **The *Daily Republic***: 'Official: self-esteem is a "social vaccine"', Kathleen L'Ecluse, 24 January 1990.
202 **The *Washington Post***: 'In Calif., Food for The Ego', Jay Matthew, 24 January 1990.
203 **The report went into reprint**: Letter from Task Force to Barbara Bush, 5 March 1990.

203 **went on to sell an extraordinary sixty thousand copies:** *The Therapeutic State: Justifying Government at Century's End*, James L. Nolan Jnr (New York University Press, 1998), Kindle location 2957.

203 **ready to 'break out across the state and country':** 'California's Newest Export', Beth Ann Krier, *LA Times*, 4 April 1990.

203 **Vasco's publicists approached *The Oprah Winfrey Show* . . . etc.:** Internal memo, 'Phase II and III of Media Plan for Maximising Release of the Final Report . . .', 19 February 1990.

204 **On 15 June the *Baltimore Sun* reported:** 'Self-esteem has arrived, says Oprah Winfrey', Mary Corey, *Baltimore Sun*, 15 June 1990.

204 **self-esteem was 'sweeping through California's public schools':** 'Self-esteem: California's newest export is a hot item', Beth Ann Krier, *Alameda Times Star*, 7 June 1990.

204 **students began meeting twice a week:** 'Self-Esteem Movement Gains Popularity in Schools', *San Francisco Chronicle*, 10 April 1990.

204 **Fifty National Council for Self Esteem chapters:** 'Self-Esteem Benchmarks: An Astonishing Revolution'. Report, Task Force to Promote Self-esteem and Personal and Social Responsibility Records, California State Archives, Sacramento, California.

204 **When Maryland launched theirs, 'phones were totally':** *The Therapeutic State: Justifying Government at Century's End*, James L. Nolan Jnr (New York University Press, 1998), Kindle location 2991.

204 **Defendants in drug trials were rewarded . . . etc.:** *The Therapeutic State: Justifying Government at Century's End*, James L. Nolan Jnr (New York University Press, 1998), Kindle location 1847.

204 **Television evangelists preached:** 'Hey, I'm Terrific!', Jerry Adler et al., *Newsweek*, 17 February 1992.

204 **kindergarten five-year-olds:** *Generation Me*, Jean Twenge (Atria, 2006), p. 55.

204 **children were awarded sports trophies:** 'Hey, I'm Terrific!', Jerry Adler et al., *Newsweek*, 17 February 1992.

204 **'It's not grade inflation':** *Nutureshock*, Ashley Merryman (Ebury, 2011), p. 115.

204 **A 1992 Gallup poll . . . police in Michigan:** 'Hey, I'm Terrific!', Jerry Adler et al., *Newsweek*, 17 February 1992.

204 **'A look at state law in all fifty':** *The Therapeutic State: Justifying Government at Century's End*, James L. Nolan Jnr (New York University Press, 1998), Kindle location 3032.

206 **'In this myth of self-esteem,' he writes:** *The Myth of Self-Esteem*, John P. Hewitt (Palgrave Macmillan, 1998), p. xiii.

215 **He'd make reference, for instance, to 'what we really all know':** These quotes were also on the 8 September 1988 tape.

215 **set them back $50,000:** Smelser on the tape: 'UC agreed to put in $50,000 to mobilize our resources to help you.'

217 **An extraordinary $30,000 was spent:** 'Advocate says self-esteem movement finally gets respect', *San Diego Union*, 22 January 1990.

217 **At its height, five publicists:** Memo to the task force from Vasconcellos RE. Blueprint for Task Force Activities, 6 March 1990.

218 **Barrowford Primary School:** My recounting of the Barrowford School story was sourced from the following: 'Headteacher whose praise for pupils went viral falls foul of Ofsted', Richard Adams, *Guardian*, 24 September 2015; 'Inspectors slam primary school where there's no such thing as a naughty child', Jata Narain, *Daily Mail*, 25 September 2015; '"There Are Many Ways Of Being Smart": Encouraging School Letter About Student Test Results Goes Viral', Megan Willett, *Business Insider*, 18 July 2014; 'School that told pupils they were better than results is rated inadequate, Ofsted', Eleanor Busby, *Times Educational Supplement*, 25 September 2015; 'School which banned teachers from raising their voices hits back at "unfair" headlines', Jon Robinson, *Lancashire Telegraph*, 7 July 2015; 'School's Letter Reminds Students That They Are More Than Just Test Scores', Rebecca Klein, *HuffPost Education*, 15 July 2014.

219 **In the town of Euclid, Cleveland:** This narrative is sourced principally from interviews with Roy Baumeister, his sister Susan and his wife, Dianne Tice.

222 **Over the course of two decades, Roy probably published:** 'Should Schools Try to Boost Self-Esteem?', Roy Baumeister, *American Educator*, Summer 1996.

223 **In 1996, Roy co-authored a review of the literature:** 'Relation of Threatened Egotism to Violence and Aggression: The Dark Side of High Self-Esteem', Roy F. Baumeister et al., *Psychological Review* (January 1996), **103**(1), pp. 5–33.

224 **He published an indignant response:** *The Art of Living Consciously*, Nathaniel Branden (Simon and Schuster, 1997), excerpt: http://www.nathanielbranden.com/what-self-esteem-is-and-is-not.

225 **Roy's study was published in May 2003:** 'Does High Self-Esteem Cause Better Performance, Interpersonal Success, Happiness or Healthier Lifestyles?', Roy Baumeister et al., *Psychological Science in the Public Interest* (May 2003), **4**(1), pp. 1–44.

225 **It's now thought that people with too much of it fail more . . . etc.:** 'Letting Go of Self-Esteem', Jennifer Crocker and Jessica J.

Carnevale, *Scientific American Mind* (September/October 2013), 24(4), pp. 26–33.

225 **In a paper co-authored with Professor Mark Leary:** 'The Nature and Function of Self-Esteem: Sociometer Theory', Mark R. Leary and Roy F. Baumeister, *Advances in Experimental Social Psychology* (2000), 22, pp. 1–62.

226 **two people playing a game in which the loser is punished:** 'Threatened Egotism, Narcissism, Self-Esteem, and Direct and Displaced Aggression: Does Self-Love or Self-Hate Lead to Violence?', Brad J. Bushman and Roy F. Baumeister, *Journal of Personality and Social Psychology* (1988), vol. 75, no. 1.

227 **It was 1999. In a basement office:** Interviews, Jean Twenge, Keith Campbell.

227–8 **in 2008, they published the details . . . that 'almost two-thirds of recent college . . . etc.':** 'Egos Inflating Over Time: A Cross-Temporal Meta-Analysis of the Narcissistic Personality Inventory', Jean M. Twenge et al., *Journal of Personality* (June 2008), 76(4), pp. 875–902.

228 **was now an 'epidemic':** *The Narcissism Epidemic*, Jean M. Twenge and W. Keith Campbell (Free Press, 2010), p. 2.

228 **rising as quickly as obesity:** *Generation Me*, Jean Twenge (Atria, 2006), p. 31.

228 **'Narcissism causes almost all of the things':** *The Narcissism Epidemic*, Jean M. Twenge and W. Keith Campbell (Free Press, 2010), p. 9.

228 **Some critics claimed that people in every generation exhibit narcissistic behaviour:** 'Every Every Every Generation Has Been the Me Me Me Generation', Elspeth Reeve, *Atlantic*, 9 May 2013; 'It Is Developmental Me, Not Generation Me: Developmental Changes Are More Important Than Generational Changes in Narcissism', Brent W. Roberts et al., *Perspectives on Psychological Science* (January 2010), 5(1), pp. 97–102.

229 **A more serious attack came from Dr Kali Trzesniewski:** 'Rethinking Generation Me: A Study of Cohort Effects From 1976–2006', Kali H. Trzesniewski and M. Brent Donnellan, *Perspectives on Psychological Science* (January 2010), 5(1), pp. 58–75; 'Reevaluating the Evidence for Increasingly Positive Self-Views among High School Students: More Evidence for Consistency Across Generations (1976–2006)', Kali H. Trzesniewski and M. Brent Donnellan, *Psychological Science* (July 2009), 20(7), pp. 920–2; 'Do Today's Young People Really Think They Are So

Extraordinary? An Examination of Secular Changes in Narcissism and Self-Enhancement', K. H. Trzesniewski, M. B. Donnellan, and R. W. Robins, *Psychological Science* (February 2008), **19**(2), pp. 181–8; 'Is "Generation Me" really more narcissistic than previous generations?', Kali H. Trzesniewski, M. Brent Donnellan, and Richard W. Robins, *Journal of Personality* (August 2008), **76**(4), pp. 903–18.

231 **an impressive 2015 study:** 'Origins of narcissism in children', Eddie Brummelman et al., *Proceedings of the National Academy of Sciences of the United States of America* (March 2015), **112**(12), pp. 3659–62.

231 **One manifestation of this is the grade inflation . . . etc.:** *Generation Me*, Jean Twenge (Atria, 2006), pp. 62, 63.

232 **In 1999, in the UK . . . etc.:** 'Record number of first-class degrees awarded to students', Josie Gurney-Read, *Daily Telegraph*, 14 January 2016; 'One in four students earns a top degree', Harry Yorke, *Daily Telegraph*, 13 January 2017.

232 **In 2012, the chief executive:** 'A-level overhaul to halt "rampant grade inflation"', Julie Henry, *Daily Telegraph*, 28 April 2012.

232 **One study, also by Twenge and Campbell . . . The wider application of school programmes:** 'Age and Birth Cohort Differences in Self-Esteem: A Cross-Temporal Meta-Analysis', Jean M. Twenge and W. Keith Campbell, *Personality and Social Psychology Review* (2001), 5(4), pp. 321–44.

233 **A 1990 story on a suddenly popular 'Self-Esteem Fair':** 'Some Bad Vibes at Self-Esteem Conference', David A. Sylvester, *San Francisco Chronicle*, 23 February 1990. The same story quotes task force member and *Chicken Soup for the Soul* author Jack Canfield coming to the same conclusion: 'There's a national explosion. The majority of my work used to be in California. The majority of my work is now outside California.'

233 **A lengthy *New Woman* article in 1991:** 'Self-Esteem: The Hope of the Future', Wanda Urbanska, *New Woman*, March 1991.

233 **A year later, a *Newsweek* cover story:** 'Hey, I'm Terrific!', Jerry Adler et al., *Newsweek*, 17 February 1992.

233 **Alan Greenspan had a problem:** My account of Greenspan's influence over Bill Clinton is largely sourced from *Maestro: Greenspan's Fed and the American Boom*, Bob Woodward (Simon and Schuster, 2000) and *Alan Greenspan: The Oracle Behind the Curtain*, E. Ray Canterbury (World Scientific, 2006).

234 **'would reach 267 miles' . . . etc.:** Clinton's Economic Plan: The

Speech; Text of the President's Address to a Joint Session of Congress; available at: http://www.nytimes.com/1993/02/18/us/clinton-s-economic-plan-speech-text-president-s-address-joint-session-congress.html?pagewanted=all.

234–5 **superstar investor Warren Buffett, of 'financial weapons of mass destruction' . . . £531tn:** 'Taking Hard New Look at a Greenspan Legacy', Peter S. Goodman, *New York Times,* 8 October 2008.

235 **In 2004 he was hailing the 'resilience':** 'Taking Hard New Look at a Greenspan Legacy', Peter S. Goodman, *New York Times,* 8 October 2008.

235 **"Where once more-marginal applicants would simply':** 'I Saw the Crisis Coming. Why Didn't the Fed?', Michael J. Burry, *New York Times,* 3 April 2010.

235 **As a result of the crash, in the US alone, more than 9 million:** 'Many Who Lost Homes to Foreclosure in Last Decade Won't Return', Laura Kusisto, *Wall Street Journal,* 20 April 2015.

235 **almost 9 million jobs disappeared:** 'Employment loss and the 2007–09 recession: an overview', Christopher J. Goodman and Steven M. Mance, *Monthly Labor Review,* April 2011.

235 **In the UK, 3.7 million people:** 'Financial Crisis Led To 3.7 Million Job Losses', http://news.sky.com/story/financial-crisis-led-to-37-million-job-losses-10454101.

235 **Researchers concluded that, in the UK, between 2008 and 2010 . . . etc.:** 'The 2008 Global Financial Crisis: effects on mental health and suicide', David Gunnell et al., University of Bristol Policy Report 3/2015.

236 **study in the *British Journal of Psychiatry*, the estimated number of additional suicides comes out at 10,000:** 'Economic suicides in the Great Recession in Europe and North America', Aaron Reeves et al., *British Journal of Psychiatry* (September 2014), **205**(3), pp. 246–7.

236 **the *New York Times* had this to say:** 'Taking Hard New Look at a Greenspan Legacy', Peter S. Goodman, *New York Times,* 8 October 2008.

236 **Many in the West have become wealthier since the 1970s:** I'm referring here to the many people whose wages have not stagnated since the start of neoliberalism, for example US employees with a bachelor's degree or a qualification greater than that. *The Rise and Fall of American Growth: The US Standard of Living Since the Civil War,* Robert J. Gordon (Princeton University Press, 2016), p. 616.

236 **Global free trade has helped lift:** *Globalisation*, Manfred B. Steger (Oxford University Press, 2013), p. 42.

236 **Around the world, the number of people living in extreme poverty:** The United Nations Millennium Development Goals Report 2015. See http://www.un.org/millenniumgoals/2015_MDG_Report/pdf/MDG%202015%20PR%20Key%20Facts%20Global.pdf.

237 **Prior to the 1940s, the wealthiest 1 per cent of the US population . . . etc.:** *Neoliberalism: A Brief History*, David Harvey (Oxford University Press, 2005), p. 15.

237 **rose to 23 per cent:** *Mirror, Mirror*, Simon Blackburn (Princeton University Press, 2014), p. 97.

237 **Between 1978 and 2014, inflation-adjusted CEO pay:** 'Top CEOs make more than 300 times the average worker', Paul Hodgson, *Fortune Magazine*, 22 June 2015.

237 **astonishing rise of the power of multinationals:** *Globalisation*, Manfred B. Steger (Oxford University Press, 2013), p. 54.

237 **median pay for the US worker:** *The Rise and Fall of American Growth: The US Standard of Living Since the Civil War*, Robert J. Gordon (Princeton University Press, 2016), p. 617.

237 **In the UK, meanwhile, between the crash and 2013, average family income:** 'Middle Income Households, 1977–2011/12', Office for National Statistics, 2 December 2013.

237 **And yet over the course of just one year, 2010–11 . . . In the UK, directors of the top 500 companies:** *Mirror, Mirror*, Simon Blackburn (Princeton University Press, 2014), p. 98.

237 **Between 1980 and 2014, US consumer credit as a percentage:** Consumer Debt Statistics, Federal Reserve, accessed at: http://www.money-zine.com/financial-planning/debt-consolidation/consumer-debt-statistics/.

237 **American students owe more than $1.2tn in college debt:** 'How The $1.2 Trillion College Debt Crisis Is Crippling Students, Parents And The Economy', Chris Denhart, *Forbes Magazine*, 7 August 2013.

237 **whilst in Britain they're £73.5bn in the red:** 'Q&A: Student loan repayments', Adam Palin, *Financial Times*, 19 June 2015.

237 **Globalization has led to cheap imports and higher levels of immigration . . . etc.:** *The Rise and Fall of American Growth: The US Standard of Living Since the Civil War*, Robert J. Gordon (Princeton University Press, 2016), p. 615.

237 **Industry is embracing harsh new working conditions:** *The Rise and*

Fall of American Growth: The US Standard of Living Since the Civil War, Robert J. Gordon (Princeton University Press, 2016), p. 614.

238 **According to the Office of National Statistics, 800,000 people:** 'Just over 800,000 people on zero-hours contract for main job', 9 March 2016, https://www.ons.gov.uk/news/news/justover800000peopleonzerohourscontractformainjob.

238 **A 2016 study found 4.5 million people, nearly one in six of all workers:** 'Nearly one in six workers in England and Wales in insecure work', Katie Allen, *Guardian*, 13 June 2016. 'Insecure work' is defined as variable shift patterns, temporary contracts, zero-hours contracts or agency contracts.

238 **In the US, meanwhile, the number of neighbourhoods classed as 'extremely poor' . . . etc.:** 'US concentrated poverty in the wake of the Great Recession', Elizabeth Kneebone and Natalie Holmes, Brookings Institute, 31 March 2016. 'Extremely poor' neighbourhoods are here defined as 'census tracts where 40 percent or more of the population lives below the federal poverty line.'

238 **Back in 1981 the American Business Roundtable:** *Mirror, Mirror*, Simon Blackburn (Princeton University Press, 2014), p. 97.

238 **In 2016, a three-year EU investigation . . . 'political crap':** 'Tim Cook condemns Apple tax ruling', Julia Kollewe, *Guardian*, 1 September 2016.

238 **some bankers in Wall Street saw their bailing out:** *This American Life*, WBEX Chicago, episode 415: 'Crybabies', originally aired 24 September 2010.

239 **As veteran tech reporter Dan Lyons has observed:** *Disrupted*, Dan Lyons (Hachette, 2016), pp. 117–18.

239 **sales of Ayn Rand's most famous work, *Atlas Shrugged*, began booming:** In the final stages of fact checking I came across an article ('Ayn Rand: the Tea Party's Miscast Matriarch', by Pam Martens, *CounterPunch*, 27 February 2012) that suggested this sales boom might have been artificially engineered by the Ayn Rand Institute, which had earmarked a large amount of money for distributing free copies. Whilst I'm sympathetic to this possibility, I'm not yet convinced the argument pans out. Two sources ('Atlas felt a sense of déjà vu', *The Economist*, 25 February 2009 and 'What Caused Atlas Shrugged Sales to Soar?,' David Boaz, *cato.org*, 18 May 2009) report that significant post-2008 sales rises of *Atlas Shrugged* registered on both Amazon.com and Neilsen's BookScan. To have this effect, the ARI would have had to have bought their copies 'over the counter' from Amazon and other book retailers, for redistribution to schools.

When I approached the ARI for comment, their publishing manager Richard E. Ralston told me, 'The increase in sales of *Atlas Shrugged* in 2009 was definitely not a result of our purchases.' He confirmed that they order and distribute a special run of editions which have a different ISBN. This means that they cannot register on BookScan or any other retailer's sales statistics.

239 **2009 saw purchases triple over 2008:** *Ayn Rand Nation*, Gary Weiss (St Martin's Griffin, 2012), p. 16.

239 **In 2011, 445,000 copies were sold:** '"Atlas Shrugged" Still Flying Off Shelves!', 14 February 2012; https://ari.aynrand.org/media-center/press-releases/2012/02/14/atlas-shrugged-still-flying-off-shelves.

239 **In 2012, Wall Street journalist Gary Weiss wrote:** *Ayn Rand Nation*, Gary Weiss (St Martin's Griffin, 2012), p. 4.

240 **nominating a secretary of state . . . etc.:** 'Ayn Rand-acolyte Donald Trump stacks his cabinet with fellow objectivists', James Hohmann, *Washington Post*, 13 December 2016. None of this is to say that Rand would've necessarily endorsed every one of Trump's statements and policies, of course. It doesn't necessarily follow that because she's influential, she'd automatically agree with the ideas and works of all the people she's influenced. In fact, as we're talking about Rand, this seems staggeringly unlikely.

Book Six: The Digital Self

page

245 **A man, Doug Engelbart, appearing in a headset:** My account of the story of Doug Engelbart, ARC, EST and Stewart Brand was mostly sourced from: *What the Dormouse Said*, John Markoff (Penguin, 2005); *From Counterculture to Cyberculture*, Fred Turner (University of Chicago Press, 2006); *The Network Revolution*, Jacques Vallee (Penguin, 1982); *Bootstrapping*, Thierry Bardini (Stanford University Press, 2000); 'Chronicle of the Death of a Laboratory: Douglas Engelbart and the Failure of the Knowledge Workshop', Thierry Bardini and Michael Friedewald, *History of Technology* (2003), **23**, pp. 191–212; 'Douglas Engelbart's lasting legacy', Tia O'Brien, *Mercury News*, 3 March 2013. A video of Engelbart's presentation is widely available online. The account I've given in the text has been lightly edited for sense and concision.

245 **audience considered Engelbart a 'crackpot':** 'The Mother of All Demos – 150 years ahead of its time', Cade Metz, *Register*, 11 December 2008.

247–8 **John Markoff has called 'a complete vision of the information age':** *What the Dormouse Said*, John Markoff (Penguin, 2005), p. 9.

249 **In 1968, the year of the demo, the Institute's co-founder Michael Murphy had written:** 'Esalen: Where Man Confronts Himself', Michael Murphy, *Stanford Alumni Almanac*, May 1968.

249 **with one 1985 *Esquire* story reporting 'scientists':** 'Encounters at the Mind's Edge', George Leonard, *Esquire*, June 1985.

249 **Tim O'Reilly, the man who in 2005 christened the internet's:** 'The Trend Spotter', Steven Levy, *Wired*, 1 October 2010.

250 **'There is an impression that Doug goes off in a corner and hatches ideas' . . . etc.:** *Bootstrapping*, Thierry Bardini (Stanford University Press, 2000), pp. 198–200.

251 **Engelbart was treating his people like 'laboratory animals':** 'Chronicle of the Death of a Laboratory: Douglas Engelbart and the Failure of the Knowledge Workshop', Thierry Bardini and Michael Friedewald, *History of Technology* (2003), **23**, pp. 191–212, at p. 206.

251 **who raced back to his company, Apple Computer:** *Steve Jobs*, Walter Isaacson (Abacus, 2015), pp. 96–7.

251 **Engelbart recalled meeting Jobs in the 1980s:** 'Douglas Engelbart's lasting legacy', Tia O'Brien, *Mercury News*, 3 March 2013.

251 **Fred Turner, would archly note that:** 'Stewart Brand's Whole Earth Catalog, the book that changed the world', Carole Cadwalladr, *Guardian*, 5 May 2013.

252 **In 1995, Brand told *Time Magazine*:** As quoted in the exhibition 'You Say You Want a Revolution? Records and Rebels 1966–1970', the Victoria and Albert Museum, London.

252 **As a student, terrified of the Communists:** *From Counterculture to Cyberculture*, Fred Turner (University of Chicago Press, 2006), pp. 41, 42.

252 **'That whole victim mindset – saying to the government':** *From Counterculture to Cyberculture*, Fred Turner (University of Chicago Press, 2006), p. 287.

253 **advising clients such as IBM, AT&T and the Pentagon:** 'Long Boom or Bust; A Leading Futurist Risks His Reputation With Ideas on Growth And High Technology', Steve Lohr, *New York Times*, 1 June 1998.

253 **an influential essay called 'The Long Boom':** 'The Long Boom: A History of the Future, 1980–2020', Peter Schwartz and Peter Leyden, *Wired*, 1 July 1997.

254 **Alan Greenspan called himself a libertarian:** 'Taking Hard New

Look at a Greenspan Legacy', Peter S. Goodman, *New York Times*, 8 October 2008.

254 **co-founded by libertarian Louis Rosetto:** *From Counterculture to Cyberculture*, Fred Turner (University of Chicago Press, 2006), p. 217.

254 **A 1994 edition covered the GBN itself:** *From Counterculture to Cyberculture*, Fred Turner (University of Chicago Press, 2006), Plate 15.

254 **'By the end of the decade,' writes Turner:** *From Counterculture to Cyberculture*, Fred Turner (University of Chicago Press, 2006), p. 215.

254 **'legitimated calls for corporate deregulation':** *From Counterculture to Cyberculture*, Fred Turner (University of Chicago Press, 2006), p. 216.

255 **Between 2006 and 2008, Facebook's user base:** *Status Update*, Alice E. Marwick (Yale University Press, 2014), p. 2.

256 **Twitter went from hosting:** https://en.wikipedia.org/wiki/Twitter.

256 **93 billion selfies . . . Every third photograph** S. Diefenbach, L. Christoforakos, 'The Selfie Paradox: Nobody Seems to Like Them Yet Everyone Has Reasons to Take Them. An Exploration of Psychological Functions of Selfies in Self-Presentation', *Frontiers in Psychology*. 2017;8:7. doi:10.3389/fpsyg.2017.00007.

257 **the average cost of a one-bedroom apartment was 23 per cent:** 'Golden gates', *The Economist*, 5 November 2015.

257 **One, the Negev, had been reported:** 'SRO tenants' tales tell scary story', Jessica Kwong, *San Francisco Examiner*, 21 November 2014.

257 **Meanwhile, Chez JJ, in the Castro:** 'An SF Hacker Hostel Faces the Real World and Loses', Davey Alba, *Wired*, 22 August 2015.

258 **The Startup Castle, a Tudor-style mansion:** 'Silicon Valley's "Startup Castle" is looking for roommates, and the requirements are completely bonkers', Kevin Roose, Fusion.net, 13 May 2015.

262 **They cite surveys that suggest . . . etc.:** *Generation Me*, Jean Twenge (Atria, 2006), p. 99.

262 **One 2006 poll of British children placed . . . Over in the US:** *The Narcissism Epidemic*, Jean M. Twenge and W. Keith Campbell (Free Press, 2010), pp. 93, 94.

262 **Twenge points to further data that suggest individualism is rising:** 'Increases in Individualistic Words and Phrases in American Books, 1960–2008', Jean M. Twenge et al., *PLOS ONE* (July 2012), 7(7); 'Fitting In or Standing Out: Trends in American Parents'

Choices for Children's Names, 1880–2007', Jean M. Twenge et al., *Social Psychological and Personality Science* (2010), 1(1), pp. 19–25.

263 **They write of the largest place of worship in America, Lakewood Church in Houston:** *The Narcissism Epidemic*, Jean M. Twenge and W. Keith Campbell (Free Press, 2010), pp. 248, 249.

269 **Steve Jobs . . . Travis Kalanick:** Wozniak on Steve Job's Resignation, Bloomberg, 24 August 2011, video, from 08:46.[Wozniak: 'He must have read some books that really were his guide in life and I think *Atlas Shrugged* might have been one that he mentioned back then.'] www.bloomberg.com/news/videos/b/d93c1b72-31e2-41de-ba4a-65a6cb4f4929. 'Silicon Valley's Most Disturbing Obsession', Nick Bilton, *Vanity Fair*, November 2016.

269 **2017 survey** David E. Broockman, Gregory Ferenstein, Neil Malhotra, 'Wealthy Elites' Policy Preferences and Economic Inequality: The Case of Technology Entrepreneurs', 5 September, 2017, https://www.gsb.stanford.edu/faculty-research/working-papers/wealthy-elites-policy-preferences-economic-inequality-case

275 **a number of working-class Democrats deserted . . . etc.:** 'The Decline of the White Working Class and the Rise of a Mass Upper Middle Class', Ruy Teixeira and Alan Abramowitz, Brookings Working Paper, April 2008.

275 **The average real income for the bottom 90 per cent:** *The Rise and Fall of American Growth: The US Standard of Living Since the Civil War*, Robert J. Gordon (Princeton University Press, 2016), p. 610.

275 **Whilst those numbers hide rises for the well educated:** The average real hourly wage for those with a college degree went up 22 per cent between 1979 and 2005. *Red, Blue and Purple America: The Future of Election Demographics*, Alan Abramowitz and Roy Teixeira (Brookings Institution Press, 2008), p. 110.

275 **During the Great Compression, those without a college education:** *The Rise and Fall of American Growth: The US Standard of Living Since the Civil War*, Robert J. Gordon (Princeton University Press, 2016), p. 609.

275 **Between 1979 and 2005:** *Red, Blue and Purple America: The Future of Election Demographics*, Alan Abramowitz and Roy Teixeira (Brookings Institution Press, 2008), p. 110.

276 **a group that shifted significantly towards the Republicans in the 2016 election:** 'Reality Check: Who voted for Donald Trump?', *BBC News*, 9 November 2016. 'There was also a big swing for voters without a high-school diploma, with Mr Trump leading 51 per cent to Mrs Clinton's 45 per cent. Four years ago, President Obama had

64 per cent support from this group compared with Mitt Romney's 35 per cent.'

276 **One analysis found that:** 'Stop Saying Trump's Win Had Nothing To Do With Economics', Ben Casselman, *fivethirtyeight.com*, 9 January 2016.

276 **She describes a powerful sense:** 'A new theory for why Trump voters are so angry – that actually makes sense', Jeff Guo, *Washington Post*, 8 November 2016.

276 **the beginning of a sharp and sustained upturn in support for smaller government:** *The Politics of Resentment*, Katherine J. Cramer (University of Chicago Press, 2016), p. 152.

276 **rise in popularity of the idea that governments should be run more like businesses:** *The Politics of Resentment*, Katherine J. Cramer (University of Chicago Press, 2016), p. 174.

277 **When political scientists compare the electorate's preferences:** *The Politics of Resentment*, Katherine J. Cramer (University of Chicago Press, 2016), p. 3.

277 **Trump said he'd 'put America first':** Trump said: 'America first will be the overriding theme of my administration.' 'Donald Trump's foreign policy: "America first"', Jeremy Diamond and Stephen Colinson, CNN, 27 April 2016.

278 **one of his closest advisers complained to reporters:** 'Ringside With Steve Bannon at Trump Tower as the President-Elect's Strategist Plots "An Entirely New Political Movement"', Michael Wolff, *Hollywood Reporter*, 18 November 2016.

278 **I'M NOT "DEPLORABLE" I'M JUST A HARD WORKING:** Spotted at: https://digest.bps.org.uk/2016/11/14/we-have-an-unfortunate-tendency-to-assume-were-morally-superior-to-others/.

279 **shopped in Macy's:** This is a detail from Professor Arlie Russel. See: 'Arlie Russell Hochschild's View of Small-Town Decay and Support for Trump', Benjamin Wallace-Wells, *New Yorker*, 20 September 2016.

280 **facilitating and accelerating the neoliberal project of globalization immensely:** *Globalisation*, Manfred B. Steger (Oxford University Press, 2013), p. 35.

280 **There are 1.7 million truck drivers in the US:** 'Robots could replace 1.7 million American truckers in the next decade', Natalie Kitroeff, *Los Angeles Times*, 25 September 2016.

280 **by 2033, nearly half of all US jobs:** 'The Future of Employment: How Susceptible are Jobs to Computerisation?', Carl Benedikt Frey

et al., 17 September 2013. Available at: www.oxfordmartin.ox.ac.uk/downloads/academic/The_Future_of_Employment.pdf.

280 **lists as a hobby** Andy Beckett, 'How Britain fell out of love with the free market', *Guardian*, 4 August 2017.

281 **A respected survey the same year** Patrick Butler, 'UK survey finds huge support for ending austerity', *Guardian*, 28 June 2017.

281 **who are among those least likely to actually use the internet:** 'Americans' Internet Access: 2000–2015', Andrew Perrin, Pew Research Center.

281 **polarization between left and right has been increasing inexorably:** *The Politics of Resentment*, Katherine J. Cramer (University of Chicago Press, 2016), p. 2.

281 **studies suggest** M. J. Crockett, 'Moral Outrage in the Digital Age', *Nature Human Behaviour* (2017) doi:10.1038/s41562-017-0213-3.

282 **One investigation found that in the final three months:** 'Viral Fake Election News Outperformed Real News On Facebook In Final Months Of The US Election', Craig Silverman, BuzzFeed, 16 November 2016.

282 **social media is 'one of our biggest problems':** 'Why social media is terrible for multiethnic democracies', Sean Illing, *Vox*, 15 November 2016.

298 **two longitudinal studies** https://www.ncbi.nlm.nih.gov/pubmed/26783723 and https://www.ncbi.nlm.nih.gov/pubmed/22268607.

298 **two hours of use** Jean Twenge, 'Making iGen's Mental Health Issues Disappear', *Psychology Today*, 31 August 2017.

299 **Not only was *The Hunger Games* partly inspired by Greek myth:** 'The Classical Roots of "The Hunger Games"', Barry Strauss, *Wall Street Journal*, 13 November 2014.

299 **nearly a hundred and fifty years** Shiv Malik, 'Was 8th June the revenge of the millennials?', *Prospect magazine*, 21 June 2017.

299 **more likely than their elders to avoid risky behaviours . . . etc.:** 'What is happening to children and young people's risk behaviours?', UK Government report available at: https://www.gov.uk/government/uploads/system/uploads/attachment_data/file/452059/Risk_behaviours_article.pdf. For the US, see the CDC's Youth Risk Behavior Surveillance System: https://www.cdc.gov/healthyyouth/data/yrbs/index.htm. Overview at https://www.cdc.gov/features/yrbs/. See also this fantastic interactive report: 'Today's teens . . .', Sarah Cliff, Soo Oh and Sarah Frostenson, *Vox*, 9 June 2016, available at http://www.vox.com/a/teens.

300 **their greater awareness of the deep structural inequalities:** I'm talking here about the rise of Identity Politics. If you need a reference for this, I envy you.

300 **some believe** Professor Keith W. Campbell, email to the author.

300 **A study of undergraduates at US universities found students more anxious:** '"I don't want to grow up, I'm a [Gen X, Y, Me] kid": Increasing maturity fears across the decades', April Smith et al., *International Journal of Behavioral Development*, 21 June 2016.

300 **In the UK, another study found four in ten young people:** 'Young Women's Trust, No Country for Young Women', Young Women's Trust Annual Survey 2016, available at: http://www.youngwomenstrust.org/assets/0000/4258/No_country_for_young_women__final_report.pdf.

300 **a 2015 Pew survey found 40 per cent of US millennials:** '40% of Millennials OK with limiting speech offensive to minorities', Jacob Poushter, Pew Research Center, 20 November 2015, available at: http://www.pewresearch.org/fact-tank/2015/11/20/40-of-millennials-ok-with-limiting-speech-offensive-to-minorities/.

300 **college students are 40 per cent less empathetic than students:** 'Changes in Dispositional Empathy in American College Students Over Time: A Meta-Analysis', Sara Konrath, *Personality and Social Psychology Review* (May 2011), 15(2), pp. 180–98.

301 **only 19 per cent of millennials believe most people can be trusted:** 'Millennials in Adulthood: Detached from Institutions, Networked with Friends', Pew Research Center, 7 March 2014, available at http://www.pewsocialtrends.org/2014/03/07/millennials-in-adulthood/.

301 **bullying at school is down:** 'Trends in Bullying and Peer Victimization', David Finkelhor, Crimes Against Children Research Center, August 2014, available at http://scholars.unh.edu/cgi/viewcontent.cgi?article=1054&context=ccrc.

301 **bullying online is rising:** 'Trends in Cyberbullying and School Bullying Victimization in a Regional Census of High School Students, 2006–2012', Kessel Schneider et al., *Journal of School Health* (September 2015), 85(9), pp. 611–20.

301 **When students at Matteo Ricci College in Seattle:** 'Under Fire, a Dean Departs', Scott Jaschik, *Inside Higher Education*, 25 July 2016; 'Supporters slam Seattle U.'s treatment of dean Jodi Kelly', Katherine Long, *Seattle Times*, 8 June 2016; 'Too many "dead white dudes"? Seattle U students protest program's curriculum', Katherine Long, *Seattle Times*, 17 May 2016.

301 **the vendor of inadequate sushi:** 'Oberlin Students Take Culture War to the Dining Hall', Katie Rogers, *New York Times*, 21 December 2015.

302 **University College London banned the Nietzsche Society:** *I Find That Offensive*, Claire Fox (Biteback, 2016), p. 48.

Book Seven: How To Stay Alive In the Age of Perfectionism

page

305 **Austen Heinz stood at the window:** My recounting of the Austen Heinz story was sourced from interviews with his sister Adrienne, his friend Mike Alfred, and (via email) Audrey Hutchinson, as well as the following: 'Interview, Sciencepreneur Extraordinaire – Pioneers Festival 2014', available at https://www.youtube.com/watch?v=QYE3ancjrjk – Pioneers Festival 2014; 'Interview, Austen Heinz invented a DNA laser printer for you to create new creatures', available at https://www.youtube.com/watch?v=cPnq5pcYfew&t=598s; 'Interview, This Week In Startups, Austen Heinz, DNA laser printing demo and exploration of the dark side', available at https://www.youtube.com/watch?v=VWMOSviTF4Y; 'Interview, Draper TV, Creating Animals from Imagination (And DNA Sequencing) | Founder of Cambrian Genomics Austen Heinz', available at https://www.youtube.com/watch?v=aiHHxnh4fsM; 'Interview, Extreme Tech Challenge, Cambrian Genomics: Democratizing Creature Creation', available at https://www.youtube.com/watch?v=D3OgIMJwdoQ; 'Interview, The Company Printing DNA Made To Order', Forbes, available at https://www.youtube.com/watch?v=petjCQDucII; 'New start-up goal: Make vaginas smell peachy (literally)', Jenny Kutner, Salon.com, 20 November 2014; 'How Not to Disrupt Women's Bodies', Maria Aspan, Slate.com, 21 November 2014.

309–13 The video of Austen's presentation was kindly provided to me by Debra Becker at DEMO. 'There's a dark side to startups, and it haunts 30% of the world's most brilliant people', Biz Carson, *Business Insider*, 1 July 2015; 'The Vagina BioHack That Wasn't: How Two "Startup Bros" Twisted and Took Credit for a Young Woman's Company', Jessica Cussins, *Biopolitical Times*, 25 November 2014; 'The CNN 10, Healing the Future', http://edition.cnn.com/interactive/2014/04/health/the-cnn-10-healing-the-future/; 'Sweet Peach won't make vaginas smell like fruit or taste of Diet Coke', Hannah Jane Parkinson, *Guardian*,

24 November 2014; 'Startup bros trying to bio-hack vaginas is the problem with Silicon Valley', Arwa Mahdawi, *Guardian*, 22 November 2014; 'The history of feminine hygiene products is far from peachy', Liz Cookman, *Guardian*, 25 November 2014; 'People with a vagina have needs – but making their baby-cannon smell of fruit is not one of them', Lindy West, *Guardian*, 25 November 2014; 'How Not to Disrupt Women's Bodies', Maria Aspan, Inc.com, 20 November 2014; 'Male Startup Founders Think Your Vagina Should Smell Like a Ripe Peach', Natasha Tiku, Valleywag, 19 November 2014; '"Sweet Peach" Probiotic Developed, Two Men Will Make Women's Vaginas Smell Like Peaches', Nina Bahadur, *Huffington Post*, 20 November 2014; 'Put Down The Pitchforks: New Product Doesn't Make Vaginas Smell Like Peaches', Chuck Bednar, Redorbit, 24 November 2014; 'R.I.P. Austen Heinz, Biotech Entrepreneur and Rebel', Zach Weissmueller, Reason.com, 30 June 2015; '"Sweet Peach has nothing to do with scent!" Female CEO of vaginal probiotic is "appalled" by male colleagues who misrepresented her product to the public', Margot Peppers, *Mail* Online, 24 November 2014; 'Sweet Peach Founder Speaks: Those Startup Dudes Were Wrong About My Company', Jeff Bercovici, Inc.com, 21 November 2014; 'Would you take a pill to smell like peaches . . . down there? Two male entrepreneurs launch probiotic supplement that alters women's natural scent', Margot Peppers, *Mail* Online, 20 November 2014; 'The Founder Of Sweet Peach Is Actually A Woman And She Doesn't Want Your Vagina To Smell Like Fruit', Arabelle Sicardi, BuzzFeed, 20 November 2014; 'Is the Sweet Peach startup a complete scam?', Selena Larson, Daily Dot Tech, 21 November 2014; 'The Vagina Bio-Hack That Wasn't: How Two "Startup Bros" Twisted the "Sweet Peach" Mission', Jessica Cussins, *Huffington Post*, 26 November 2014; 'These 2 tech bros want to make vaginas smell like peaches', Selena Larson, Daily Dot Tech, 19 November 2014; 'These Startup Dudes Want to Make Women's Private Parts Smell Like Ripe Fruit', Jeff Bercovici, Inc.com, 19 November 2014; 'Why We Need to Talk More About Mental Illness in Tech and Business', Jeff Bercovici, Inc.com, 7 July 2015; *Life Without a Windshield*, Austen James (Broken Science, 2009).

317 **'Oh, Mom! Dad! I'd like you to meet Professor Little'**: Interview, Brian Little.

318 **'beyond doubt that neuroticism causes awful, private':** *Personality*, Daniel Nettle (Oxford University Press, 2009), p. 243.

318 **'it is so closely associated with it that it is hard':** *Personality*, Daniel Nettle (Oxford University Press, 2009), p. 114.

318 **'the worried actually have more to worry about':** *Personality*, Daniel Nettle (Oxford University Press, 2009), p. 120.

318 **'Seductive though it might be':** *Personality*, Daniel Nettle (Oxford University Press, 2009), p. 103.

321 **therapeutic interventions can affect:** *Personality Psychology*, R. Larsen et al. (McGraw Hill, 2013), p. 115. See also 'Volitional Personality Trait Change: Can People Choose to Change Their Personality Traits?', Nathan W. Hudson and R. Chris Fraley, *Journal of Personality and Social Psychology* (September 2015), **109**(3), pp. 490–507. Just as this book was going to press, a new study ('A Systematic Review of Personality Trait Change Through Intervention', Brent Roberts et al., *Psychological Bulletin*, 5 January 2017) was published that found therapeutic interventions could have relatively large effects on neuroticism and much more moderate effects on other traits. Included in those therapeutic interventions were SSRIs and other drugs. Part of the effect seemed to be patients recovering from the crisis that sent them to seek help in the first place. One of the authors, Professor Brent Roberts, explained, 'We're not saying personality dramatically reorganizes itself. You're not taking an introvert and making them into an extravert. But this [study] reveals that personality does develop and it can be developed.' (Thanks to Dr Stuart Ritchie for his emergency assistance with this!) https://news.illinois.edu/blog/view/6367/448720.

321 **decrease in openness as we exit middle age:** *Personality Psychology*, R. Larsen et al. (McGraw Hill, 2013), p. 114, fig. 5.3.

321 **Take agreeableness. Imagine how many interactions you have every week:** For more on this idea, see *Personality*, Daniel Nettle (Oxford University Press, 2009), p. 43.

322 **Through the processes of 'social evocation' and 'social selection' . . . Our personalities also trigger responses:** *Personality*, Daniel Nettle (Oxford University Press, 2009), p. 46.

323 **Extraverts really do talk more:** *Personality*, Daniel Nettle (Oxford University Press, 2009), p. 46.

323 **high neuroticism is a strong predictor of divorce:** 'Personality and compatibility: a prospective analysis of marital stability and marital

satisfaction', E. L. Kelly and J. J. Conley, *Journal of Personality and Social Psychology* (January 1987), 52(1), pp. 27–40.

323 **risk of death, in any given year, by around 30 per cent:** 'Psychosocial and behavioral predictors of longevity: The aging and death of the "Termites."', H. S. Friedman et al., *American Psychologist* (February 1995), 50(2), pp. 69–78.

323 **extraversion is a predictor of promiscuity:** 'The relationship between personality traits and sexual variety seeking', Bita Nasrollahi et al., *Procedia – Social and Behavioral Sciences* (2011), **30**, pp. 1399–402.

323 **A major study that tracked the lives of six hundred US men:** *The Natural History of Alcoholism Revisited*, George E. Valliant (Harvard University Press, 1995).

323 **they're the euphoric drunks:** *Me, Myself and Why*, Jennifer Ouellette (Penguin, 2014), pp. 132–4.

323 **A person's genotype also changes how they're affected:** Interview, Daniel Nettle.

323 **Conscientiousness, too, is positively correlated with prosperity:** 'What grades and achievement tests measure', Lex Borghans et al., *Proceedings of the National Academy of Sciences of the United States of America* (November 2016), 113(47), pp. 13354–9.

325 **Some sceptics of these ideas point to the work that shows we can become different people in different contexts:** See, for example, *The Self Illusion*, Bruce Hood (Constable, 2011), pp. 173–4.

325 **'free traits' . . . Writes Little, 'a biogenically agreeable woman:** *Me, Myself and Us*, Brian Little (PublicAffairs, 2014), p. 62.

331 **in her book *Quiet* Susan Cain has famously described the 'extrovert ideal':** *Quiet*, Susan Cain (Penguin, 2013), Part One: The Extrovert Ideal.

333 **'highly active people who can lay their hands on:** *Personality*, Daniel Nettle (Oxford University Press, 2009), pp. 83, 84.

Index